SpringerBriefs in Operations Research

For further volumes:
http://www.springer.com/series/11467

Michael Doumpos · Constantin Zopounidis

Multicriteria Analysis in Finance

Springer

Michael Doumpos
School of Production Engineering
and Management
Technical University of Crete
Chania
Greece

Constantin Zopounidis
School of Production Engineering
and Management
Technical University of Crete
Chania
Greece

and

Audencia Nantes School of Management
Nantes
France

ISSN 2195-0482 ISSN 2195-0504 (electronic)
ISBN 978-3-319-05863-4 ISBN 978-3-319-05864-1 (eBook)
DOI 10.1007/978-3-319-05864-1
Springer Cham Heidelberg New York Dordrecht London

Library of Congress Control Number: 2014935977

Printed on acid-free paper

Springer is part of Springer Science+Business Media (www.springer.com)

Preface

Since the 1970s, the field of finance has evolved rapidly, driven by the advances in information technology and the introduction of financial innovations involving new financial products and services. Nowadays, investors have a wide range of options suitable for different investment policies (e.g., equity, different types of funds, fixed income products, derivatives, etc.), managers of firms use a variety of products for corporate financing and risk management, and policy makers face new challenges in choosing the best policies and measures for monitoring and controlling the markets in an effective way. Despite the significant progress that has been achieved, in today's globalized and increasingly volatile environment, modern financial theory still faces a number of challenging issues, including but not limited to:

- the management of different types of financial risks,
- understanding of the factors that affect the global markets,
- analysis of the performance of firms and organizations,
- regulatory issues related to the implementation of effective supervision practices.

In this context, the decision-making process in the "new era" of finance is becoming more and more difficult, thus making the development and implementation of effective operational decision support tools a critical requirement. Toward this direction multidisciplinary, integrated, and operational approaches are needed to cover the complex and multidimensional nature of the financial decision-making process.

To this end, three different levels of analysis can be considered. The first is focused on building models that describe the characteristics of financial problems and phenomena. Mainstream areas such as financial economics, mathematical finance, and financial engineering employ stochastics, probability theory, statistical and optimization models, among others, for financial modeling purposes. At the second level, empirical studies are required to test the theoretical models and identify new unobserved explanatory factors. This is usually done through statistical and econometric methods that facilitate the analysis of panel and cross-sectional financial data related to the operation of financial markets and the actions of firms, investors, and policy makers. Finally, the third level of analysis is related to the construction of possible solutions and the selection of the most appropriate ways of action. Originating from the work of Markowitz on portfolio selection, optimization methods have been widely used for formulating solutions in many financial decision problems.

At the decision-making level, it is now widely acknowledged that financial decisions require the consideration of multiple factors, variables, and criteria, in a framework that needs to be flexible and customizable to the requirements of a particular situation. Financial decision makers combine statistical estimates and forecasts, domain knowledge derived from the theory of finance, and constraints imposed by the external environment, with their own expertise, judgments, and decision-making policy. In this process, multiple perspectives, goals, and decision criteria are involved. Financial modeling is often based on the assumption that financial decisions are driven by a wealth maximization objective, but this single objective is often not well-defined, thus requiring a broader description through multiple subobjectives or alternative factors.

Multiple criteria decision aid (MCDA) is well suited in this context. MCDA provides a wide range of analytic methodological tools for decision aiding under multiple conflicting criteria and it is particularly well suited for financial decision support. MCDA contributes at several levels of the financial decision-making process, covering both the problem structuring stages and algorithmic issues related to the construction and assessment of satisfactory solutions.

This book intends to provide a comprehensive overview of the applications of MCDA approaches to financial decisions. The book is organized into seven chapters. Chapter 1 starts with an introduction to the main aspects of the financial decision-making process, including an overview of different types of financial decisions and a discussion of the connections between financial modeling, risk management, and financial engineering. The multicriteria aspects of financial decisions are also analyzed in detail, from different perspectives. Chapter 2 covers the founding principles, main concepts, and techniques in the area of MCDA, including multiobjective optimization, multiattribute value theory, outranking relations, and disaggregation methods. The remaining chapters Chaps. 3–6 focus on particular areas of financial decisions, including banking, credit granting, portfolio, management, investment appraisal, and country risk assessment. Each chapter presents the contributions and applications of different MCDA methods in these areas. Illustrative applications are also presented to demonstrate the applicability and results of MCDA methods. The book closes with a discussion of some important open issues that pose challenges for the future development of the MCDA paradigm and its application in financial decision aiding.

Chania, Greece, February 2014

Michael Doumpos
Constantin Zopounidis

Contents

Chapter 1
Introduction to Financial Decision Making

Abstract Financial decision problems are ill-structured and involve complex big data of uncertain nature. This chapter analyzes the multiple aspects of financial decisions and discusses their main features. The contributions of financial modeling and financial engineering are also discussed from the perspective of risk management, which has been a core issue in the recent developments in the field. Finally, we elaborate on the multidimensional aspects of financial decisions considering different perspectives.

Keywords Financial decisions · Financial modeling · Risk management · Financial engineering

1.1 The Nature of Financial Decisions

Financial theory and practice is involved with a wide range of topics related to the financial operation of firms and organizations as well as the choices and behavior of institutional and individual investors. Some typical examples include capital budgeting, asset valuation and management, financial planning, corporate financing, and investment appraisal, among others.

Financial decisions can be classified in two broad main categories, namely investment and financing decisions [34, 84] as illustrated in Fig. 1.1. Investment decisions are involved with the appraisal of investment projects and real assets, the allocation of capital to them, and the management of existing investments. Long-term investments fall in the area of capital budgeting [13], whereas short-term ones (e.g, inventories of raw materials and finished products, cash, deposits, and receivables) are related to working capital management. On the other hand, financing decisions are concerned with the ways that funds can be obtained to finance long-term investments and day-to-day operating needs. The available options include different types of debt and equity. In a corporate setting, the chosen mix of these two options defines the capital

M. Doumpos and C. Zopounidis, *Multicriteria Analysis in Finance*,
SpringerBriefs in Operations Research, DOI: 10.1007/978-3-319-05864-1_1,

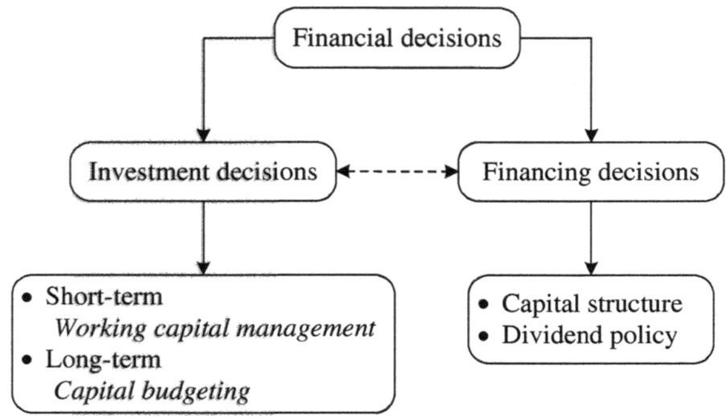

Fig. 1.1 A categorization of financial decisions

structure of the firm and its dividend policy. It should be noted at this point that investment and financing decisions often interact, thus requiring their consideration in an integrated manner [84].

The context in which financial decisions are taken has changed drastically over the past decades. The globalization of the financial and business environment, along with the increasing importance of financial markets have led to a number of major innovations regarding the available financial products and services. These provide financial managers, investors, and policy makers with new capabilities, but also raise new challenges. For instance, investors now have a wide range of investment options, ranging from private equity, different types of funds, fixed income securities (e.g., bonds), commodities, etc. Similarly, the financing of firms and decisions related to their optimal capital structure have become much more involved. Funds can be raised from different sources (e.g., bank financing, equity markets, bonds, structured products, venture capital, etc.), new issues emerge related to corporate governance, and the effect of externalities has become stronger, as demonstrated by the global financial crisis of 2007–2008. On the other hand, policy makers are concerned with monitoring and regulating the operation of financial institutions and financial markets, with an emphasis on ensuring long-term financial stability. Their policy decisions and actions have a direct impact on all areas of financial services and the sentiment in the markets, thus also having indirect effects on the financial decisions in other business sectors (manufacturing, commerce, non-financial services, etc.), as well as individuals.

In this context, some important features characterizing financial decision problems can be highlighted:

- *Risk and uncertainty*: Financial decisions are characterized by high uncertainties, which makes risk a core concept and risk modeling a critical process. In the next section we shall elaborate on the importance of risk management in financial decision making.
- *Imprecision and unpredictability*: A significant part of financial risks is simply due to the fact that financial decisions are taken in a socio-economic context driven by human behavior (collective and individual), which is impossible to describe and explain analytically, without some level of ambiguity. As a consequence, the

underlying relationships that govern the "financial world" and the broader socio-economic environment, are highly complex and (by definition) incomplete. Thus, financial problems have inherit imprecisions, which make financial phenomena difficult to analyze and predict with the level of accuracy often found in natural sciences and engineering. This may seem troublesome, but as argued by van der Wijst [255], it is in fact completely natural and well in line with the fundamental assumption of finance theory about the efficiency of financial markets.

- *Dynamic and static problems*: Financial decision problems combine dynamic and static aspects. For instance, managing the risk of asset portfolios, requires the consideration of the dynamics of the assets' values over time. On the other hand, a bank's decision to accept or reject a credit application is mostly static as it is based on a static estimate of the applicant's likelihood of default for the next time period (usually one year).
- *Strategic and operational decisions*: Financial decisions are taken at both the strategic and the operational level. The former involves decisions of non-repetitive nature with long-term effects (e.g., a merger or acquisition), which require analyses customized to each particular decision instance. Operational decisions, on the other hand, are taken regularly and usually require timely results (often in real-time; e.g., active trading of financial assets).
- *Big data*: With the advances in information technology, there is now available a vast volume of financial data regarding financial transactions, corporate information (financial and non-financial), as well as macroeconomic data and estimates. Exploiting such data to derive non-trivial managerial information about trends and relationships involving financial factors through data-driven approaches in a predictive and explanatory framework, provides the potential for greatly enhancing the financial decision making process. This is, however, a challenging issue as the data are by definition noisy and incomplete, given the aforementioned natural imprecision that characterizes finance.
- *Discrete and continuous problems*: Discrete problems refer to cases where a finite set of options should be analyzed and evaluated. Typical examples in finance include investment appraisal, credit scoring/rating, country risk analysis, etc. Continuous problems, on the other hand, involve cases where the available options can only be implicitly described by a set of requirements they should meet, but cannot be enumerated exhaustively (e.g., portfolio optimization and asset-liability management).

In addition to the above characteristics, financial decisions are multi-faceted and can be viewed from multiple perspectives. This can be illustrated by considering the example of mean-variance analysis for optimal portfolio management. The decision maker's problem is how to form a combination of risky assets such that, for a given level of expected return, risk is minimal. From a decision theory point of view this is an expected quadratic utility maximization problem. From a probabilistic point of view it is seen as the returns being generated by elliptical distributions. From an operations research point of view it can be seen as a constrained quadratic programming problem. Further, from an econometrics point of view it involves the prediction of expected returns and the covariance matrix.

In the next section we discuss the different aspects of financial models and their uses for financial engineering and risk management.

1.2 Financial Modeling, Risk Management, and Financial Engineering

The diverse, multi-facet, and complex nature of financial decisions cannot be properly addressed without developing and implementing proper modeling approaches based on analytic techniques. These provide the necessary tools to integrate all the ingredients for the problem through a structured approach. Financial models are formal representations of the relationships that describe a financial system of interest. Such representations enable analysts and decision makers to understand the underlying structure of a problem, identify its critical aspects, test hypotheses, as well as to construct and evaluate alternative ways of actions. According to Spronk and Hallerbach [224] "financial modeling is concerned with the development of tools supporting firms, investors, intermediaries, governments, etc. in their financial-economic decision making, including the validation of the premises behind these tools and the measurement of the efficacy of these tools".

The modeling approaches used in the context of finance have become considerably sophisticated over the past decades. Markowitz [167] first introduced analytic quantitative techniques for portfolio selection and risk management, based on optimization and utility theory. Later, the publication of the Black and Scholes option pricing model [29] set the basis of the modern era of finance, which is characterized by the adoption of a much more analytic-engineering approach.

Financial models have different purposes and forms. In particular, following a decision theory framework [22], three main categories can be defined:

- The first category of models focuses on normative theories, which describe how "optimal" decisions should be made assuming rational and well-informed decision makers. Such models are based on assumptions on market efficiency and norms of rational behavior as prescribed by the expected utility theory of von Neumann and Morgenstern [185]. The Black-Scholes option pricing model and the capital asset pricing model [216] are typical examples in this category.
- The second class of financial models is motivated by empirical evidence indicating that actual decisions often do not follow prescribed norms and theories of rational behavior [20, 198]. Thus, models are developed to *describe* how actual decisions are taken (descriptive models). Similar to normative models, descriptive ones are also general, but are developed on empirical data involving the actions and decisions of economic agents and decision makers in practice. Behavioral finance [19] is involved with such types of models and research questions. The foundations of this approach have been set by the introduction of prospect theory [137], which asserts that decisions are based on changes in wealth combined through subjective

decision weights rather than the level of wealth in uncertain outcomes aggregated through probabilities, as assumed in expected utility theory.
- Normative and descriptive models set the basis for understanding the operation of the financial system, which defines the context in which financial decisions are taken. However, they do not provide direct guide on what actions should be implemented in a specific problem instance. Prescriptive models fill in this gap, by supporting financial decision makers in their daily practice. Their objective is to facilitate the decision process taking into account the specific characteristics of each particular instance, in combination with the general knowledge that normative and descriptive models provide.

The realization of such financial models is based on a variety of quantitative techniques from different disciplines. For instance:

- stochastic models are used for the valuation of financial derivatives and modeling the price and volatility dynamics of financial assets [218],
- statistical and econometric methods are employed to build models that explain and predict asset prices and analyze the actions of firms, investors, and regulators, using time series, panel, and/or cross-sectional data [210],
- probability theory and simulation are used for risk assessment [102],
- optimization models are used for constructing asset allocation strategies [49],
- computational intelligence methods are used for discovering complex patterns in financial data for predictive decision making and for handling computationally intensive large-scale problems [76, 101].

Financial models haven become increasingly sophisticated over the years, and have led to a major transformation of the field of finance to a science with strong analytic as well as engineering focus [168]. Financial engineering first appeared as a term in the 1980s to describe this transformation that began with the introduction of the Black-Scholes option pricing formula. Finnetry [90] described financial engineering as the design, development and implementation of innovative financial instruments and processes and the formulation of innovative solutions to financial decision making problems. Marshall and Dorigan [168] refer to a similar definition given by the International Association of Financial Engineers, which defines financial engineering as "the development and creative application of financial technology to solve financial problems".[1]

Similarly to traditional engineering disciplines, financial engineering uses domain knowledge to support the design of new (financial) instruments and process, thus leading to financial innovations [153, 246], which are essential for reducing the cost of capital, cutting down operating costs, and ensuring diversification of risks [60]. Financial instruments range from traditional products such as special savings accounts, bonds, stocks, different types of funds, to financial derivatives (e.g., options, futures, swaps, etc.), as well as structured instruments such as asset-backed securities. Instruments are combined with solutions and processes, which define the steps required to

[1] http://www.iaqf.org/senior-fellows

tackle a given problem and the actions needed to implement solutions successfully in the context of firm's complex operation.

Mulvey et al. [183] illustrate some typical uses of financial engineering in four major areas of financial management:

- in corporate finance, it is used to minimize the cost of the funds required for the operation of a firm or to engineer takeovers and buyouts,
- in securities trading, it provides the tools to develop dynamic trading strategies,
- in investment management, it supports the design and development of new investment products (e.g., high yield mutual funds), or the transformation of high-risk investments into low risk ones, and
- it introduces risk management strategies, such as asset-liability management, hedging, and portfolio insurance.

Nevertheless, the rapid growth of financial engineering and the resulting widespread use of complex financial instruments have been heavily criticized for their role in the outbreak of the 2007–2008 global crisis [60, 91]. Among the many aspects of this criticism, the risk management practices followed in the financial sector and the regulatory framework on risk management prior to the crisis, have attracted much attention.

Of course, financial risk management is not a new issue. However, the frequent turmoils in the global markets have brought financial risk management into the spotlight and the efforts made by regulators and policy makers to ensure long-term financial stability have intensified. The introduction of the 1988 Basel Accord was the first regulatory attempt to impose common global rules on the risk management practices of financial institutions. The revised framework of Basel II first published in 2004 brought major changes based on a more refined analysis of financial risks, together with stricter requirements, and more involved procedures, whereas the third update (Basel III) is already under development.

Financial risk management is a complex process that aims towards minimizing the undesirable effects due to adverse conditions, that may arise from the variability in asset prices, the internal operation of the firms, and their relation with other entities. However, it should be emphasized that risk management does ensure that risk is eliminated, as risks are always present. With that in mind, financial risk management is multi-stage process requiring at least the following three main activities, which are implemented in a dynamic setting [56, 118]:

1. Identifying the relevant key financial risks that should be managed and setting risk tolerance levels and priorities.
2. Assessing the risks associated with the relevant activities.
3. Designing and implementing risk control and reduction strategies.

Financial risks can have different forms. In accordance with the Basel regulatory framework [18], we can identify four main categories of financial risks faced by firms and organizations, as shown in Table 1.1.

Table 1.1 Main categories of financial risks

Credit risk	Default
	Counterparty risk
	Country risk
Operational risk	Human factors
	Systems failures
	Processes
	External events
Liquidity risk	Funding liquidity
	Market liquidity
Market risk	Equity prices
	Foreign exchange
	Interest rates
	Commodity prices

- *Credit risk* originates from the inability of borrowers to meet their debt obligations towards their creditors. This includes a borrower's internal default risk (idiosyncratic risk), the connections of a borrower with other counterparties (i.e., clients, suppliers, debtors, etc.; counterparty risk), and a systematic component related to country (sovereign) risk (i.e., the risk that a country will default).
- *Operational risk* refers to the internal operation of a firm related to risks that arise due to human errors and malpractice (e.g., fraud), the failure of technical systems inside the firm, inadequate processes (e.g., internal control procedures), and adverse external events (e.g., political and social turmoil, changes in the regulatory framework, natural disasters).
- *Liquidity risk* arises when a financial transaction cannot be conducted or when early liquidation is required in order to meet payments obligations. This can be due to mismatching incoming flows and scheduled payments (funding liquidity risk) or to limited market liquidity.
- *Market risk* is the systematic risk derived from adverse movements in financial markets, which cannot be reduced by diversification strategies. Among others, it involves equity prices, exchange rates, interest rates, and commodity prices.

This categorization demonstrates that in an integrated risk management framework, different risks should be considered. This raises the question of whether multiple measures of risk should be used. Furthermore, the risk-return trade-off that characterizes all financial decisions should also be taken into consideration. Beginning from such observations, the following section outlines a new paradigm for financial decisions based on the premise that different conflicting objectives should be meet. This is illustrated through the analysis of the multiple facets of corporate objectives, including return and risk.

1.3 The Multicriteria Aspects of Financial Decisions

1.3.1 Corporate Objectives

The finance theory has adopted the principle of wealth maximization as the single objective that drives decision making in the corporate world. Adopting this principle provides financial managers a single target that guides the decision making process. Furthermore, it is also easy to monitor and evaluate the results of the decisions taken, usually on the basis of risk-adjusted performance measurement approaches.

However, this approach may be too simplistic to describe the daily operation of the business environment. Bhaskar and McNamee [26] note that even if a firm does focus on a single objective, this is often way too broadly defined, thus requiring the introduction of multiple proxy goals that can be translated to everyday terms. Steuer and Na [230] also argue that wealth maximization is not understood in a common way by all stakeholders, as they often have different conceptions of wealth, risk, liquidity, social responsibility, environmental protection, employee welfare, etc. Hallerbach and Spronk [111] on the other hand, emphasize the role of the imperfections in the decision environment, such as information asymmetries, conflicting interests, and transactions costs, which restrict the available opportunities and require a much richer description. Furthermore, it should be noted that financial decisions are also relevant to nonprofit entities, public organizations, regulators, and policy makers, whose decisions and actions are inevitably driven by a set of complex socio-economic principles.

From a financial modeling perspective, adopting a normative or descriptive approach [225] with a single performance measure is a mathematically convenient approach. For instance:

- Financial planning models are often built assuming a wealth maximization objective with policy and risk constraints. However, transforming goals into constraints alters the nature of the problem, making it difficult for the decision maker to explore in a comprehensive way the trade-offs that may be involved among multiple goals and parameters. Identifying, analyzing, and measuring such trade-offs provides very useful insights for taking more informed financial decisions.
- In a different context, models analyzing and describing corporate performance are often based on proxies of the overall financial performance of the firms. For instance, there are numerous studies exploring the factors that best describe the stock market returns of a firm, its profitability (usually measured by the return on assets), or its growth (e.g., sales and profit growth). The results from such empirical studies are important for the understanding of what drives such success factors (i.e., market performance, profitability, growth etc.). However, the evaluation of corporate financial performance requires the adoption of a holistic approach combining all the relevant factors, including financial criteria (profitability, leverage, liquidity, solvency, managerial performance) and non-financial information which is crucial for the viability of a corporate entity (i.e., organizational structure, market position, competitive advantages, management competence, etc.).

Thus, the "traditional" perspective of financial theory is indeed useful for understanding the operation of the financial markets, the activities of firms and organizations, and the identification of relevant decision factors. Nevertheless, a realistic approach should be directed towards facilitating decision makers in the consideration of all pertinent decision criteria, the analysis of the trade-offs involved, the suggestion of multiple alternatives ways of actions, and their evaluation. Such a comprehensive framework based on multiple criteria may not be straightforward to implement. However, it acts both as a holistic modeling approach for financial decisions, as well as a tool for exploring non-trivial aspects related to the problem, constructing solutions, and facilitating their implementation, thus becoming a learning tool for financial decision makers.

1.3.2 The Multidimensional Aspects of Risk

The traditional mean-variance framework of Markowitz [167] considers risk as the variance of assets' returns. In his 1959 book, Markowitz also briefly discussed extensions involving other risk measures (e.g., semi-variance, mean absolute deviation, expected loss) and highlighted the importance of combining statistical estimates with the expert judgment of portfolio analysts.

Since the 1990s the research on the introduction of proper risk measures has grown rapidly, focusing mainly on downside risk measures, which emphasize on the risk of loss, rather than the variance of returns. The introduction of value at risk [135] and its widespread adoption by practitioners and regulators has added to the debate on suitable ways for risk assessment, mainly due to its drawbacks and unappealing properties [9, 235, 236].

It should also be emphasized that not only there are multiple risk measures and factors, but also different perceptions of risk and risk attitudes. In a normative context, specific utility functions (of wealth) are assumed to model risk aversion. For instance, Markowitz's mean-variance model implicitly assumes a quadratic utility function. However, a decision maker's attitude towards risk is inevitably subjective and it is connected to the utility of the alternatives under consideration [47]. As a consequence, general risk models grounded on financial and economic principles should be combined with operational techniques providing individualized decision support in the context of a specific financial problem and the risk attitude of a particular decision maker.

1.3.3 Return, Profitability, and Wealth

Return is usually easy to conceptualize, but its accurate estimation and prediction are quite challenging. Researchers and practitioners devote much effort in predicting future returns and identifying factors that support the selection of profitable investments and ways of actions. In the case of financial assets, multi-factor models have become the standard in this area. The best known example is arbitrage pricing theory [205], whereas newer popular extensions and variants can be found is several works such as those of Fama and French [85, 86] and Carhart [40].

When dealing with financial assets return is simply the rate of change in the assets' value. In different settings, however, such when considering corporate performance, return and the relevant concept of profitability are not defined through a single measure. Generally, return and profitability in such cases refer to the economic results of a firm's activities compared to the resources used to achieve them. This general definition leads to multiple profitability measures covering different economic results and resources. For instance, in assessing corporate profitability, managers, analysts, and investors take into consideration criteria such as the shareholders' return (return on equity), the way that corporate assets are used in a profitable way (return on assets), the cost structure of the firm (e.g., profit margins), as well as the ability of the firm to maintain a stream of strong future cash flows. Recent developments add further complexities, highlighting new aspects of long-term profitability, including among others the principles of corporate social responsibility, sustainability, and socially responsible investments. Thus, the traditional financial principle of wealth maximization is actually much more involved than a well-defined, single-objective framework [225, 230].

Chapter 2
An Overview of Multiple Criteria Decision Aid

Abstract This chapter provides an overview of the multicriteria decision aid paradigm. The discussion covers the main features and concepts in the field as well as an introduction to the main methodological approaches and techniques.

Keywords Multicriteria decision aid · Multiobjective optimization · Goal programming · Multiattribute value theory · Outranking relations · Preference disaggregation analysis

2.1 Introduction

The discussion in the previous chapter highlights the multi-dimensional nature of financial decisions. When it comes to actual decision support, the particular characteristics of the decision environment in a given instance should be considered, together with the preferences and judgment policy of the decision makers, and the domain knowledge provided by normative and descriptive financial theories. For instance, one should consider specific budgetary, regulatory, or policy constraints and conditions, qualitative expert judgments describing special aspects of the problem, and other special features which are relevant for a given decision context.

Thus, the complex and often ill-defined nature of important concepts such as risk and return, the multiple explanatory and decision factors involved, as well as the complex framework in which financial decisions are taken and implemented, calls for integrated decision aid tools. These should support the structuring of the problem, the modeling process, the identification and evaluation of alternative ways of action, as well as the implementation of the selected solutions. Operations research and management science (OR/MS) techniques address these issues and provide a wide range of modeling tools, suitable for handling financial decision problems under different schemes with regard to the decision context, the available information, and data.

M. Doumpos and C. Zopounidis, *Multicriteria Analysis in Finance*,
SpringerBriefs in Operations Research, DOI: 10.1007/978-3-319-05864-1_2,

Multiple-criteria decision aid (MCDA) has evolved over the past decades becoming a major discipline in OR/MS. The field of MCDA is devoted to the development and implementation of decision support tools and methodologies for facilitating decision making in ill-structured problems involving conflicting multiple criteria, goals, objectives, and points of view.

In the context of MCDA a wide variety of decision settings can be considered, including among others static deterministic problems, decisions under uncertainty and fuzziness, dynamic problems, as well as group decision making. In all cases, the MCDA paradigm is based on the comprehensive description of a particular decision problem taking into account all the pertinent decision factors, on the basis of the decision makers' preferences. This is an appealing approach in many domains, including finance, given the high complexity that characterize the decisions that firms and organizations take and the multiple points of view which are involved (financial, regulatory, social, environmental, etc.). The following sections provide a brief overview of the MCDA field. An comprehensive introduction to the main concepts, principles and techniques in this field can be found in the book of Belton and Stewart [24], whereas the recent advances and research trends are presented in the books of Ehrgott et al. [80], and Zopounidis and Pardalos [270].

2.2 Main Concepts of Multicriteria Analysis

The main goal of MCDA is to provide decision aiding in complex and ill-structured problems, in accordance with the decision makers' preferential system and judgment policy. When multiple decision criteria are involved, there cannot be a unanimous optimal decision (in the traditional optimization sense), as different goals and objectives naturally lead to the formulation of different recommendations. However, having formal procedures and analytic techniques for problem structuring and the assessment of alternative ways of action, greatly facilitates the decision process.

MCDA intervenes in all phases of the decision process, beginning from problem structuring [24, 256] up to the implementation of the recommended solutions. An outline of the decision aiding process in the context of MCDA is illustrated in Fig. 2.1, following the approach introduced by Roy [207].

The first level of the above process, involves the specification of a set $\mathscr{A}$ of feasible alternative solutions to the problem at hand (alternatives). The objective of the decision is also determined. The set $\mathscr{A}$ can be continuous or discrete. In the former case $\mathscr{A}$ is specified through constraints imposed by the decision maker or by the decision environment. In the case where $\mathscr{A}$ is discrete, a finite set of alternatives are subject to evaluation.

The determination of the objective of the decision specifies the way that $\mathscr{A}$ should be considered to take the final decision. This involves the selection of the decision problematic that is most suitable to the problem at hand:

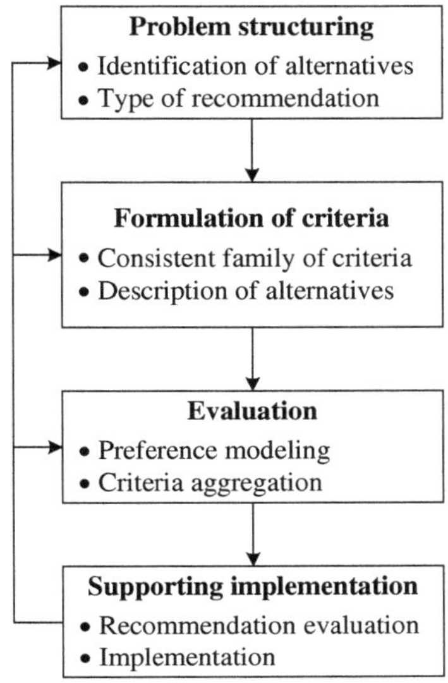

Fig. 2.1 The decision aiding process in MCDA

- Choice of the best alternative(s).
- Ranking of the alternatives from the best to the worst.
- Classification/sorting of the alternatives into pre-defined performance categories.
- Description of the alternatives.

The selection of an investment project is a typical choice example, whereas the ranking of a bank's branches on the basis of their efficiency and performance is an example of a ranking problem. Financial decisions that require a classification of the available options include credit scoring (see Chap. 4), mergers and acquisitions (e.g., identification of firms that could be takeover targets), and country risk assessment (see Chap. 6), among others. Finally, the descriptive problematic may involve the identification of alternatives with similar performance characteristics (firms, investments, countries, etc.). It must be noted, however, that often a combination of different problematics is required in order to address a given problem instance. For instance, in Sect. 3.3 we shall analyze a case study regarding the development of a bank rating methodology that uses a ranking scheme to define a classification of banks into performance categories.

The second stage involves the identification of all factors related to the decision. MCDA assumes that these factors have the form of criteria. A criterion is a real function f measuring the performance of the alternatives on each of their individual characteristics, defined such that:

$$f(\mathbf{x}) > f(\mathbf{y}) \Leftrightarrow \mathbf{x} \succ \mathbf{y} \text{ (alternative } \mathbf{x} \text{ is preferred over alternative } \mathbf{y})$$
$$f(\mathbf{x}) = f(\mathbf{y}) \Leftrightarrow \mathbf{x} \sim \mathbf{y} \text{ (alternatives } \mathbf{x} \text{ and } \mathbf{y} \text{ are indifferent)}$$

The set of the criteria $F = \{f_1, f_2, \ldots f_K\}$ identified at this second stage of the decision aiding process, must form a consistent family of criteria. A consistent family of criteria is characterized by the following properties [31]:

- Monotonicity: If alternative $\mathbf{x}$ is preferred over alternative $\mathbf{y}$, the same should also hold for any alternative $\mathbf{z}$ such that $f_k(\mathbf{z}) \geq f_k(\mathbf{x})$ for all k.
- Completeness: If $f_k(\mathbf{x}) = f_k(\mathbf{y})$ for all criteria, then the decision maker should be indifferent between alternatives $\mathbf{x}$ and $\mathbf{y}$.
- Non-redundancy: The set of criteria satisfies the non-redundancy property if the elimination of any criterion results to the violation of monotonicity and/or completeness.

Once a consistent family of criteria has been specified, the next step is to proceed with the specification of the criteria aggregation model that meets the requirements of the problem. Finally, the last stage involves all the necessary supportive actions needed for the successful implementation of the results of the analysis and the justification of the model's recommendations.

2.3 Methodological Approaches

MCDA provides a wide range of methodologies for addressing decision-making problems of different types. The differences between these methodologies involve the form of the models, the model development process, and their scope of application. On the basis of these characteristics, the following four main streams in MCDA research can be distinguished [194]:

- Multiobjective optimization.
- Multiattribute utility/value theory.
- Outranking relations.
- Preference disaggregation analysis.

The following sections provide a brief overview of these methodological streams.

2.3.1 Multiobjective Optimization

Multiobjective optimization (MOO) extends the traditional single optimization framework to problems with multiple objectives. Formally, a MOO problem has the following form:

$$\begin{array}{ll} \max & f_1(\mathbf{x}), f_2(\mathbf{x}), \ldots, f_K(\mathbf{x}) \\ \text{subject to: } \mathbf{x} \in \mathscr{A} & \end{array} \tag{2.1}$$

where $\mathbf{x}$ is the vector of decision variables, $f_1, f_2, \ldots, f_K$ are the objective functions (all assumed to be in maximization form) and $\mathscr{A}$ is the set of feasible solutions. The objectives are assumed to be in conflict, which implies that they are not all optimized simultaneously at a single solution.

In this context, optimality is defined on the basis of the concept of dominance. A feasible solution $\mathbf{x}^*$ dominates another solution $\mathbf{x} \in \mathscr{A}$ if and only if $f_k(\mathbf{x}^*) \geq f_k(\mathbf{x}), \forall k = 1, \ldots, K$, with at least one of the inequalities being strict. Thus, solving problem (2.1) requires the identification of Pareto optimal solutions, that is solutions not dominated by others.

The identification of the set of Pareto optimal solutions can be done with several techniques. A comprehensive discussion of various algorithmic procedures and formulations can be found in the books of Miettinen [179] and Steuer [229]. For instance, a commonly used approach is based on aggregating the objectives through a scalarization function. A typical example is the Chebyshev scalarization model, which is suitable for both convex and non-convex problem instances [229]. The model can be expressed as follows:

$$\begin{aligned} \min \quad & \gamma + \rho \sum_{k=1}^{K} [f_k^* - f_k(\mathbf{x})] \\ \text{s.t.} \quad & \gamma \geq w_k [f_k^* - f_k(\mathbf{x})] \\ & \mathbf{x} \in \mathscr{A}, \gamma \geq 0 \end{aligned} \tag{2.2}$$

where w_k is the non-negative trade-off constant for objective k, f_k^* is the maximum value of objective k (which can be found by performing K single objective optimizations), and ρ is a small positive constant used to exclude the possibility of obtaining weakly efficient solutions.[1] The full set of efficient (Pareto) solutions can be traced by solving the above optimization problem with different trade-offs for the objectives.

MOO problems can also be expressed in the form of goal programming (GP) formulations. In a GP context, the decision maker specifies target levels $t_1, t_2, \ldots, t_K$ for the objectives. A GP model can be expressed in the following general form:

$$\begin{aligned} \min \quad & D(d_k^+, d_k^-) \\ \text{subject to:} \quad & f_k(\mathbf{x}) + g_k(d_k^+, d_k^-) \leq = \geq t_k, \quad k = 1, 2, \ldots, K \\ & \mathbf{x} \in \mathscr{A} \\ & d_k^+, d_k^- \geq 0, \quad k = 1, 2, \ldots, K \end{aligned} \tag{2.3}$$

where d_k^+, d_k^- are slack variables indicating the deviations from the pre-specified target levels, whereas D and $g_1, \ldots, g_K$ are functions of the slack variables. The first set of constraints defines the relationship between the objectives, the associated target levels, and the slack variables.

For instance, a goal of the form "objective k should be approximately equal to t_k" can be formulated as $f_k(\mathbf{x}) + d_k^+ - d_k^- = t_k$, with $d_k^+ + d_k^-$ being minimized. Similarly, a goal of the form "objective k should be at least equal to t_k, if possible" is formulated as $f_j(\mathbf{x}) + d_k^+ \geq t_k$, such that d_k^+ is minimized. Following the same

[1] A feasible solution $\mathbf{x}^*$ is called weakly efficient if there is no other feasible solution $\mathbf{x}$ such that $f_k(\mathbf{x}) > f_k(\mathbf{x}^*)$, for all $k = 1, \ldots, K$.

approach, different types of goals can be introduced in the general model (2.3). An detailed analysis of GP models and their applications can be found in the book of Jones and Tamiz [134].

2.3.2 Multiattribute Value Theory

Utility theory has played a central role in the field of decision analysis since its axiomatization by von Neumann and Morgenstern [185]. In a multicriteria context, multiattribute utility/value theory (MAUT/MAVT)[2] provides a normative approach for characterizing and analyzing rational decision making [139]. MAVT is mostly involved with the way decision makers make choices among a finite set of alternatives, but it also has important implications for MOO and GP models [78].

In particular, MAVT is involved with functional decision models (utility functions) aggregating multiple criteria into a composite indicator. A value function V aggregates a vector $\mathbf{x}$ of K decision criteria such that:

$$V(\mathbf{x}_i) > V(\mathbf{x}_j) \Rightarrow \text{alternative } i \text{ is preferred over alternative } j \ (\mathbf{x}_i \succ \mathbf{x}_j)$$
$$V(\mathbf{x}_i) = V(\mathbf{x}_j) \Rightarrow \text{alternatives } i \text{ and } j \text{ are indifferent } (\mathbf{x}_i \sim \mathbf{x}_j)$$

Depending on the criteria independence conditions, different form of value functions can be defined. For instance, if it is assumed that the preferences of the decision maker on any subset of criteria do not depend on the other criteria (mutual preferential independence), then V is expressed in additive form:

$$V(\mathbf{x}) = \sum_{k=1}^{K} w_k v_k(x_k)$$

where $w_k \geq 0$ is the trade-off constant for criterion k and $v_k(x_k)$ is the associated marginal value function. This is a compensatory model, in the sense that the low performance in one criterion can be compensated by a high performance on others. The trade-off constants define this level of compensation. For instance, a poor performance on a criterion that has a high trade-off constant is not easily compensated by high performance by other criteria with low trade-offs. The trade-offs are by definition non-negative and they are usually normalized such that they sum up to a predefined scaling constant (e.g., $w_1 + w_2 + \cdots + w_K = 1$). On the other hand, the marginal value functions decompose the overall performance score into partial scores at the criteria level; they are non-decreasing for criteria in maximization form (e.g., profit related criteria) and non-increasing for minimization criteria (e.g.,

[2] The term "utility theory" is usually used in the context of decisions under uncertainty, whereas "value theory" is often preferred for deterministic problems. Having this distinction in mind, in order to simplify the presentation in the remainder of the book we shall use the term "value" to cover both situations.

risk criteria). Similarly to the trade-offs, the marginal values are also appropriately scaled, usually between 0 (for the worst performing alternative) and 1 (for the best performance).

Under weaker preferential independence assumptions alternative value models can be introduced. For instance, a multiplicative value function is expressed as follows:

$$1 + \lambda V(\mathbf{x}) = \prod_{k=1}^{K} [1 + \lambda w_k v_k(x_k)]$$

where $\lambda > -1$ is a scaling constant, such that $1 + \lambda = \prod_{k=1}^{K} [1 + \lambda w_k]$. In the case $\lambda = 0$ the multiplicative function reduces to an additive one.

Under the more general setting, the multilinear value function can be considered:

$$\begin{aligned} V(\mathbf{x}) &= \sum_{k=1}^{K} w_k v_k(x_k) + \sum_{k=1}^{K} \sum_{\ell > k} w_{k\ell} v_k(x_k) v_\ell(x_\ell) \\ &+ \sum_{k=1}^{K} \sum_{\ell > k} \sum_{z > \ell} w_{k\ell z} v_k(x_k) v_\ell(x_\ell) v_z(x_z) + \cdots \\ &+ w_{123\ldots} v_1(x_1) v_2(x_2) v_3(x_3) \cdots \end{aligned}$$

This general model has $2^K - 1$ scaling constants as opposed to K trade-offs involved in the additive and multiplicative forms, and includes these two simpler models as special cases. However, the additional complexity of the multilinear model makes it difficult to use in cases with $K \geq 4$. Nevertheless, Keeney and Raiffa [139] note that even when their underlying assumptions do not hold, additive and multiplicative are reasonable approximations to the general case.

2.3.3 Outranking Relations

The founding principles of outranking techniques can be traced to social choice theory [8]. An operational framework in the context of decision aiding, was first introduced by Roy [208] with the ELECTRE methods (ELimination Et Choix Traduisant la REalité).

In contrast to the functional models employed in the context of MAVT, outranking models are expressed in relational form through which the validity of affirmations such as "alternative i is at least as good as (or preferred over) alternative j" can be analyzed. Exploiting such pairwise comparisons through appropriate procedures leads to the final evaluation results (i.e., choice of the best ways of action, ranking or classification of finite set of alternatives from the best to the worst ones).

For instance, in the context of the ELECTRE methods [89] the evaluation process is based on pairwise comparisons used to assess the strength of the outranking relation "alternative i is at least as good as alternative j" ($\mathbf{x}_i \, S \, \mathbf{x}_j$). The comparisons are performed at two stages. The first involves the concordance test, in which the strength of the indications supporting the outranking relation is assessed. This can be done through the following concordance index:

$$C(\mathbf{x}_i, \mathbf{x}_j) = \sum_{k=1}^{K} w_k c_k(x_{ik}, x_{jk})$$

where x_{ik} and x_{jk} are the data for the two alternatives on criterion k, w_k is the weight (relative importance) of criterion k, and $c_k(x_{ik}, x_{jk})$ is the criterion's partial concordance index, defined such that:

$$c_k(x_{ik}, x_{jk}) = \begin{cases} 0 & \text{if } x_{ik} < x_{jk} - p_k \\ \frac{x_{ik} - x_{jk} + p_k}{p_k - q_k} & \text{if } x_{jk} - p_k \leq x_{ik} \leq x_{jk} - q_k \\ 1 & \text{if } x_{ik} > x_{jk} - q_k \end{cases}$$

where p_k and q_k are the user-defined preference and indifference thresholds for criterion k ($p_k \geq q_k \geq 0$). The case $C(\mathbf{x}_i, \mathbf{x}_j) = 1$ indicates that the outranking relation is clearly verified by all performance criteria, whereas the case $C(\mathbf{x}_i, \mathbf{x}_j) = 0$ indicates that there is no evidence to support the hypothesis that alternative i outranks alternative j.

At the second stage, the strength of the indications against the outranking relation is assessed through the calculation of a discordance index for each criterion:

$$d_k(x_{ik}, x_{jk}) = \begin{cases} 0 & \text{if } x_{ik} > x_{jk} - p_k \\ \frac{x_{ik} - x_{jk} + p_k}{p_k - v_k} & \text{if } x_{jk} - v_k \leq x_{ik} \leq x_{jk} - v_k \\ 1 & \text{if } x_{ik} < x_{jk} - v_k \end{cases}$$

The discordance indices examine the existence of veto conditions, in which the performance of alternative i may be too low in one or more criteria (i.e., $d_k(x_{ik}, x_{jk}) \approx 1$ for some k) and consequently it cannot be concluded that it outranks alternative j, irrespective of its performance on the rest of the evaluation factors. The veto threshold $v_k \geq p_k$ defines the minimum difference $x_{jk} - x_{ik}$ above which veto applies according to the performances of alternatives i and j on criterion k.

The combination of the two stages can be performed in different ways. For example, in the ELECTRE III method the following credibility index is used:

$$\sigma(\mathbf{x}_i, \mathbf{x}_j) = C(\mathbf{x}_i, \mathbf{x}_j) \prod_{k \in \mathscr{F}} \frac{1 - d_k(x_{ik}, x_{jk})}{1 - C(\mathbf{x}_i, \mathbf{x}_j)}$$

where $\mathscr{F}$ is the set of performance criteria such that $d_k(x_{ik}, x_{jk}) > C(\mathbf{x}_i, \mathbf{x}_j)$. Credibility indices close to one indicate that the outranking relation $\mathbf{x}_i \, S \, \mathbf{x}_j$ is almost surely true, whereas $\sigma(\mathbf{x}_i, \mathbf{x}_j) \approx 0$ indicates that the relation cannot be verified. On the basis of the results of such pairwise tests, different procedures can be used to chose the best alternatives, or to rank and classify them into categories.

Some particular special features of outranking models, include the consideration of non-compensatory and intransitive preferences. Non-compensation enriches the traditional preference and indifference relations, through the modeling of incomparability. Incomparability arises in situations where alternatives with special characteristics are considered (e.g., excellent performance on some criteria, but very poor performance on others). In such cases it may be difficult to derive straightforward conclusions on the overall performance of the alternatives. On the other hand, handling intransitive preference structures enables the modeling of situations where for example $\mathbf{x} \succ \mathbf{y}$ and $\mathbf{y} \succ \mathbf{z}$ does not imply $\mathbf{x} \succ \mathbf{z}$.

These features of outranking methods are particularly well-suited for financial decision making. For instance, the non-compensatory character of outranking models fits well the emphasis that finance practitioners and policy makers put on minimizing downside risk. With non-compensation, particularly risky features of the available options are identified and their trade-offs with other performance criteria are eliminated. On the other hand, cases where intransitivity and incomparability arise should be examined more closely with additional analysis possibly focusing on qualitative factors. For instance, Doumpos and Zopounidis [74] found that, in the context of corporate credit scoring, such cases are most likely to arise for firms whose financial characteristics are not enough to formulate an accurate recommendation.

Apart from the ELECTRE methods, the PROMETHEE [33] methods have also been widely used for building and exploiting outranking/preference relations in decision aiding. An example of using such an evaluation technique in the context of banking management is given in Sect. 3.3.1. An overview of other outranking techniques can be found in [169].

2.3.4 Preference Disaggregation Analysis

The development of the decision models in MCDA is based on direct and indirect procedures. Direct procedures require the decision maker to specify the parameters of the model (e.g., the criteria trade-offs) through interactive, structured communication sessions in cooperation with the decision analyst. In some cases this might be a feasible process, mainly when the decision involves strategic choices of non-repetitive character. In other cases, however, where real-time decision making is required, such direct procedures are not applicable. Furthermore, the cognitive difficulties associated with direct elicitation procedures, are also an important factor. Indirect preference disaggregation methods are very helpful in this context [128]. Preference disaggregation analysis (PDA) uses regression-like techniques to infer a decision model from a set of decision examples on some reference alternatives, so that the model is as consistent as possible with the actual evaluation of the alternatives by the decision maker.

The key assumption in PDA is that the decision maker is unable or unwilling to provide direct information about his/her system of preferences, other than a sample of decisions that he/she has taken in the past or would take in a given future situation. Given this set of sample decisions (reference set), the analyst should provide the decision maker with a starting basis upon which he/she can elaborate on the specific details of his/her preferential system. In this context, inferring a model that is consistent with the given sample decisions can be of great help to the decision aiding process.

The reference set is the main input in a PDA process; it may consist of past decisions, a subset of the alternatives under consideration, or a set of fictitious alternatives which can be easily judged by the decision maker [128]. Depending on the decision problematic, the evaluation of the reference alternatives may be expressed by defining an order structure (total, weak, partial, etc.) or by classifying them into appropriate classes.

Formally, let $Y(X')$ denote the decision maker's evaluation of a set X' consisting of M reference alternatives described over K criteria. Such an evaluation may involve a complete or partial ranking of the alternatives or their classification in predefined categories and it is assumed to be based (implicitly) on a decision model $F(\mathbf{x}; \boldsymbol{\alpha})$ defined by some parameters $\boldsymbol{\alpha}$, which represent the actual preferential system of the decision maker. The objective of PDA is to infer the "optimal" parameters $\widehat{\boldsymbol{\alpha}}^*$ that approximate, as accurately as possible, the actual preferential system of the decision maker as represented in the unknown set of parameters $\boldsymbol{\alpha}$, i.e.:

$$\widehat{\boldsymbol{\alpha}}^* = \arg\min_{\widehat{\boldsymbol{\alpha}} \in \mathscr{A}} \|\widehat{\boldsymbol{\alpha}} - \boldsymbol{\alpha}\| \tag{2.4}$$

where $\mathscr{A}$ is a set of feasible values for the parameters $\widehat{\boldsymbol{\alpha}}$. With the obtained parameters, the evaluations performed with the corresponding decision model $F(\mathbf{x}; \widehat{\boldsymbol{\alpha}}^*)$ will be consistent with the evaluations actually performed by the decision maker for any set of alternatives.

Problem (2.4), however, cannot be solved explicitly because $\boldsymbol{\alpha}$ is unknown. Instead, an empirical estimation approach is employed using the decision maker's evaluation of the reference alternatives to proxy $\boldsymbol{\alpha}$. Thus, the general form of the optimization problem is expressed as follows:

$$\widehat{\boldsymbol{\alpha}}^* = \arg\min_{\widehat{\boldsymbol{\alpha}} \in \mathscr{A}} L[Y(X'), \widehat{Y}(X')] \tag{2.5}$$

where $\widehat{Y}(X')$ denotes the recommendations of the model for the alternatives in X' and $L(\cdot)$ is a function that measures the differences between $Y(X')$ and $\widehat{Y}(X')$.

For instance, consider an ordinal regression setting, where a decision maker wants to construct a model for ranking some alternatives from the best to the worst ones. The decision maker has evaluated the six alternatives of Table 2.1 under three criteria and provided a ranking (last column) from the best ($y_i = 1$) to the worst one ($y_i = 6$).

The decision maker has decided to construct a decision model of the form $V(\mathbf{x}) = w_1x_1 + w_2x_2 + w_3x_3$, such that $w_1, w_2, w_3 \geq 0$ and $w_1 + w_2 + w_3 = 1$, which is as consistent as possible with the provided evaluations. In order to be consistent with the information in the given reference set, the model should satisfy the inequality

Table 2.1 A reference set for constructing a ordinal regression model

Alternatives	Criteria			Ranking
	x_1	x_2	x_3	(y)
$\mathbf{x}_1$	7	1	8	1
$\mathbf{x}_2$	4	5	8	2
$\mathbf{x}_3$	10	4	2	3
$\mathbf{x}_4$	2	4	1	4
$\mathbf{x}_5$	4	1	1	5
$\mathbf{x}_6$	1	2	5	6

$V(\mathbf{x}_i) > V(\mathbf{x}_j)$, for all pairs of alternatives such that $y_i < y_j$ (i.e., alternative i is preferred over alternative j, $\mathbf{x}_i \succ \mathbf{x}_j$). Such a model can be constructed through the solution of the following linear program:

$$
\begin{aligned}
\min \quad & \sum_{i=1}^{6} \sum_{j \neq i} \varepsilon_{ij} \\
\text{s.t.} \quad & \sum_{k=1}^{3} w_k (x_{ik} - x_{jk}) + \varepsilon_{ij} \geq \delta \quad \forall \mathbf{x}_i \succ \mathbf{x}_j \\
& w_1 + w_2 + w_3 = 1 \\
& w_k, \varepsilon_{ij} \geq 0 \qquad\qquad \forall i, j, k
\end{aligned}
$$

where $\varepsilon_{ij} = \max\{0, V(\mathbf{x}_j) - V(\mathbf{x}_i)\}$ is the absolute error for the pair of alternatives $\mathbf{x}_i \succ \mathbf{x}_j$ and δ is a small positive constant.

The foundations of PDA have been set during the 1950s with the introduction of non-parametric regression techniques using goal-programming formulations [254] and their later extension to ordinal regression models [226]. Jacquet-Lagrèze and Siskos [127] first defined the PDA framework in the context of decision aiding through the introduction of the UTA method, which is based on an additive utility modeling approach. However, other decision models can also be employed, including non-linear utility functions [37], rule-based models [105], outranking models [67, 182], Choquet integrals [104], and kernel models [191].

A comprehensive bibliography on preference disaggregation methods can be found in Jacquet-Lagrèze and Siskos [128], whereas some recent trends are discussed in [220].

Chapter 3
Banking Management

Abstract Banking institutions play a central role in the financial and business environment. Decision making in banking involves a wide spectrum of issues. This chapter focuses on the evaluation of the performance of banks. In this context, a multicriteria approach is presented, which is based on an outranking method. The proposed multicriteria methodology is illustrated through an application on a sample of Greek banks. The relationship with bank efficiency assessment is also discussed.

Keywords Banking · Bank risk rating · Outranking methods · Bank efficiency

3.1 The Regulatory Framework

Banks are at the heart of the worldwide financial system, acting as intermediaries by providing credit to firms and individuals using deposits and their investment activities. Over the years, the role of banks has undergone significant changes and their importance has increased. Nowadays, banks have extended their range of traditional commercial activities, through the introduction of specialized deposit, financing and investment products, providing new services to their customers, and expanding their operations in the global financial markets. Clearly, this context creates a wide range of new opportunities. At the same time, however, it also creates a plethora of challenges, as it has been clearly demonstrated by the recent credit crisis that began from the USA and later transmitted to Europe in the form of a banking and sovereign debt crisis.

As a consequence of the diverse nature of a bank's operation, the area of banking management is involved with a wide range of issues related to all types of financial risks faced by banks, their investment and financing activities, the efficiency of their operation, as well as the regulatory and supervisory framework that governs their full range of operations. The latter has been a focal point for policy makers over the past two decades.

M. Doumpos and C. Zopounidis, *Multicriteria Analysis in Finance*,
SpringerBriefs in Operations Research, DOI: 10.1007/978-3-319-05864-1_3,

The regulatory framework of Basel II [18], which is currently active, has been designed to improve the risk management practices in financial institutions and ensure the stability of the global financial system. The framework consists of three mutually related pillars.

- The first pillar (minimum capital requirements) is related to procedures required for specifying the minimum level of capital, which must be reserved by financial institutions, as a safety net against the undertaken risks.
- The second pillar (supervisory review process) defines the procedures that must be adopted (a) from the supervisors in order to evaluate how well banks are assessing their capital needs and (b) from the bank's risk managers for ensuring that the bank has reserved a sufficient capital to support its risks.
- Finally, the third pillar (market discipline) requires from banks to provide disclosures with how senior management and the board of directors assess and manage the various types of risk.

The upcoming revision of Basel III is expected to bring a more refined approach with new risk dimensions (e.g., liquidity risk). Even through it is now apparent that the existing regulatory framework failed to prevent the global credit crunch of 2007–2008, the adoption of common rules in a global context can be indeed positive for financial stability.

3.2 Bank Performance Evaluation

Obviously the implementation of successful policies at all levels of a bank's operation should lead to improved overall performance and reduced exposure to excessive risks. The evaluation of the performance and viability of banks has received much interest among researchers, bank managers, and regulators. Such evaluations are performed considering all factors that describe the activities, operations, and risks of a bank. The most popular evaluation framework is based on the consideration of multiple performance and risk attributes categorized in six major dimensions:

1. capital adequacy,
2. assets quality,
3. management competence,
4. earnings generating ability,
5. liquidity, and
6. sensitivity to market risks.

The evaluation context consisting of these dimensions is known as CAMELS (capital, assets, management, earnings, liquidity, sensitivity to market risks). Due to lack of sufficient historical data about bank defaults, bank performance evaluation systems are usually based on empirical assessment techniques (i.e., peer assessments). Sahajwala and Van den Bergh [212] present a comprehensive overview of the practices followed by supervisory authorities in G10 countries with respect to the adoption of risk assessment and early warning systems used for evaluating and monitoring the performance of banks. The overview indicates that central banks often use more than one system based on CAMELS and other similar frameworks, usually following a peer review approach combining financial and qualitative data.

The diverse nature of the evaluation criteria involved (qualitative and quantitative) as well as the importance of incorporating the judgments of expert banking analysts, makes MCDA a well-suited approach for building bank evaluation models. Several multicriteria techniques have been used in this context. For instance, Mareschal and Brans [163], Mareschal and Mertens [164] as well as Kosmidou and Zopounidis [146] used the PROMETHEE method, Zopounidis et al. [264] and Spathis et al. [223] used disaggregation techniques, Raveh [200] used the Co-plot method, whereas Ho [117] implemented the grey relational analysis. The evaluation of banking institutions has also been explored in a ranking context using goal programming formulations inspired by data envelopment analysis [11, 100, 126], which is a common technique for efficiency analysis with numerous applications in banking (see [88] for a comprehensive review).

Most of these studies have focused on the financial aspects of the performance of banks, using financial criteria mainly in the form of financial ratios. Other studies using MCDA approaches have considered additional aspects related to the regulatory and supervisory framework [95, 96, 125], customer-oriented criteria [106, 227]), while specific pillars of the Basel II capital adequacy framework have also been considered (e.g., operational risk [21].

Of course, banking management is not restricted to bank performance evaluation. Other important areas with applications of multicriteria techniques include:

- Asset-liability management [57, 145, 238].
- Bank branches network management [126].
- Evaluation of electronic banking services [122, 201].
- Customer relationship management [106].

The rest of this chapter is devoted to the presentation of a multicriteria approach for bank risk rating. The proposed methodology is based on the PROMETHEE II method and it has implemented in a decision support system currently, which is currently used by the Bank of Greece [73].

3.3 A Multicriteria Approach for Bank Risk Rating

The main output of bank rating models is an evaluation of the overall risk and performance of banks. In a supervisory context, expert analysts (supervisors of a central bank) gather detailed information that enables the evaluation of a bank's condition and the monitoring of its compliance with the regulatory framework. The result of this evaluation process is a rating (CAMELS rating), which provides a forward-looking approach of a bank's current overall condition and potential risk.

In common practice, the ratings are usually assigned in a scale of 1–5, which *resembles* an ordinal classification setting. Banks with ratings of 1 or 2 are considered to present few supervisory concerns, while banks with higher ratings present moderate to extreme degrees of supervisory concern. The definition of the grades in such a rating system, is based on the composite score of the banks obtained by aggregating their performance on all evaluation criteria. This score is expressed on a scale similar to the ratings (e.g., in [1, 5] or [0.5, 5.5]) so that each rating can be matched to a predefined score interval. Within this context, bank rating does not correspond to a "traditional" multicriteria classification problem, in the sense that the actual outcome of the evaluation process is a numerical evaluation score, which is matched to a risk grade at the final stage of the evaluation process, as a means of "defuzzification". This approach provides flexibility to the supervisory authorities, which may take similar actions for banks whose rating scores are very similar, even if they correspond to different ratings.

In accordance, with the CAMELS model which is currently in use by the Bank of Greece, a multicriteria methodology has been implemented that enables not only to define the required risk grades, but also to develop an overall performance index that permits comparisons on the relative performance of the banks. The methodology is based on the PROMETHEE II method [33]. The workflow of the methodology is given in Fig. 3.1.

The PROMETHEE method is widely used to rank a set of alternatives on the basis of pairwise comparisons. Except for this kind of analysis, the method was also used to perform an absolute evaluation in comparison to a pre-specified reference point. Thus, the use of the PROMETHEE method enables the consideration of both the relative and absolute performance of the banks in a unified context. The relative evaluation enables the consideration of the strengths and weaknesses of a bank as opposed to other banks (i.e., on the basis of the conditions that prevail in the banking sector), whereas the absolute evaluation enables the analysis of the condition of a bank compared to predefined reference points representing specific risk profiles. The combination of these approaches provides supervisors with a comprehensive view of the risks that banks face, taking into account the characteristics of each individual bank, the interrelationships between the banks, and the overall condition of the banking sector. The consideration of these two issues in other MCDA models (e.g., a value function) would require the introduction of specific criteria, which were difficult to define and measure in this case.

The subsections below provide details on the implementation of the PROMETHEE method in both these contexts. Details on the evaluation criteria and the details of the evaluation process are given in Sect. 3.4.

3.3.1 Relative Evaluation

The evaluation of the banks in the context of the PROMETHEE method is based on pairwise comparisons. In particular, for each pair of banks (i, j) the global preference

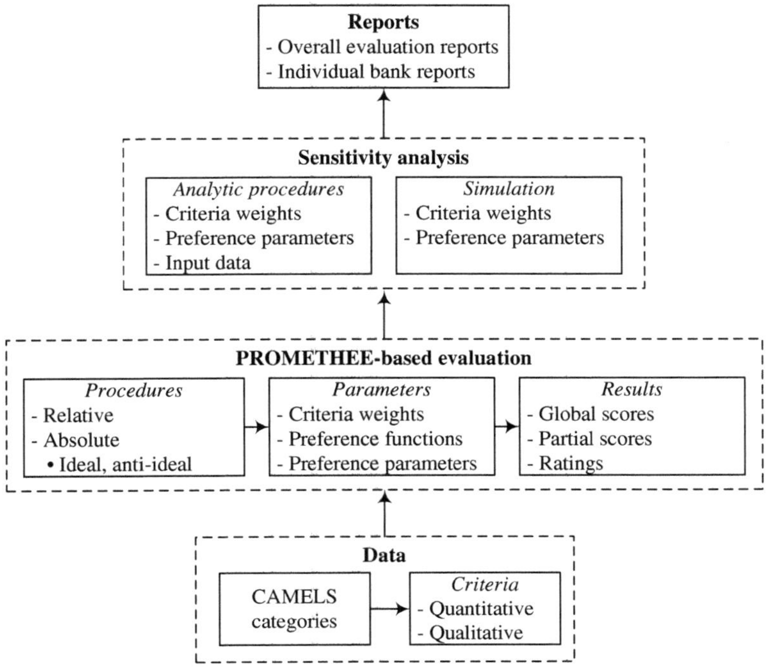

Fig. 3.1 The workflow of the multicriteria methodology

index $P(\mathbf{x}_i, \mathbf{x}_j)$ is computed, where $\mathbf{x}_i = (x_{i1}, x_{i2}, \dots, x_{in})$ is the vector with the description of bank i on n evaluation criteria. The global preference index is defined as the weighted sum of partial preference indices:

$$P(\mathbf{x}_i, \mathbf{x}_j) = \sum_{k=1}^{n} w_k \pi_k(x_{ik}, x_{jk})$$

where w_k is the weight of criterion k and $\pi_k(x_{ik}, x_{jk})$ is the corresponding partial preference index, which measures (in a [0, 1] scale) the strength of the preference for bank i over bank j on criterion k.

The partial preference index $\pi_k(x_{ik}, x_{jk})$ is a function of the difference $x_{ik} - x_{jk}$ in the performances of the banks on criterion k. A popular choice is the Gaussian function:

$$\pi_k(x_{ik}, x_{jk}) = \begin{cases} 0 & \text{if } x_{ik} \leq x_{jk} \\ 1 - \exp\left[-\dfrac{(x_{ik} - x_{jk})^2}{2\sigma_k^2}\right] & \text{if } x_{ik} > x_{jk} \end{cases}$$

where $\sigma_k > 0$ is a user defined parameter. If a low value is used for σ_k, then even a small difference $x_{ik} - x_{jk} > 0$ may lead to a significant preference for bank i over bank j. On the contrary, for large values of σ_k, strict preference may only occur when $x_{ik} \gg x_{jk}$.

An alternative function for the definition of the partial preference index is the linear generalized criterion:

$$\pi_k(x_{ik}, x_{jk}) = \begin{cases} 0 & \text{if } x_{ik} - x_{jk} \leq 0 \\ \dfrac{x_{ik} - x_{jk}}{p_k} & \text{if } 0 < x_{ik} - x_{jk} \leq p_k \\ 1 & \text{if } x_{ik} - x_{jk} > p_k \end{cases}$$

where $p_k > 0$ is the preference threshold, which defines the minimum difference $x_{ik} - x_{jk}$ above which bank i is assumed to be strictly preferred over bank j on criterion k. Note that the above functions are only meaningful for quantitative data, but alternative options for handling qualitative criteria [33].

Assuming a set of M banks under evaluation, the results of all the pairwise comparisons are aggregated into a global performance index (net flow) as follows:

$$\Phi(\mathbf{x}_i) = \frac{1}{M-1}\left[\phi^+(\mathbf{x}_i) - \phi^-(\mathbf{x}_i)\right] \tag{3.1}$$

where $\phi^+(\mathbf{x}_i) = \sum_{j \neq i} P(\mathbf{x}_i, \mathbf{x}_j)$ is the outgoing flow representing the outranking character of bank i over all the other banks and $\phi^-(\mathbf{x}_i) = \sum_{j \neq i} P(\mathbf{x}_j, \mathbf{x}_i)$ is the incoming flow representing the outranking character of all banks in the sample over bank i. Thus, the above net flow index combines the strengths and weaknesses of a bank compared to its competitors in an overall evaluation measure. The overall net flow index $\Phi(\mathbf{x}_i)$ ranges in $[-1, 1]$, with higher values associated with low risk/high performance banks.

The net flow index (3.1) can be alternatively written in additive form as:

$$\Phi(\mathbf{x}_i) = \sum_{k=1}^{K} w_k \phi_k(\mathbf{x}_i) \tag{3.2}$$

where $\phi_k(\mathbf{x}_i) = \varphi_k^+(\mathbf{x}_i) - \varphi_k^-(\mathbf{x}_i)$ is the partial evaluation score (uni-criterion net flow), defined for criterion k, with

$$\varphi_k^+(x_{ik}) = \frac{1}{M-1}\sum_{j \neq i} \pi_k(x_{ik}, x_{jk}) \quad \text{and} \quad \varphi_k^-(x_{ik}) = \frac{1}{M-1}\sum_{j \neq i} \pi_k(x_{jk}, x_{ik})$$

representing, respectively, the strengths and weaknesses of bank i compared to the others with respect to criterion k.

The advantage of using the additive form (3.2) over (3.1) is that it provides a decomposition of the overall performance of a bank on each evaluation criterion through the corresponding uni-criterion flows. Thus, the strengths and weaknesses of the bank can be easily identified in terms of the criteria.

In order to build the required bank rating model, the evaluation scale for both the overall performance index, as well as for all the partial performance indices can be modified to enable the definition of a 5-point rating scale. In this model calibration step, the partial net flows $\phi_k(\mathbf{x}_i)$ can be used to define a modified partial evaluation function as follows:

$$v_k(x_{ik}) = \begin{cases} 0.5 & \text{if } x_{ik} \geq x_k^* \\ 0.5 + 5\dfrac{\phi_k(x_{ik}) - \phi_k(x_k^*)}{\phi_k(x_{k*}) - \phi_k(x_k^*)} & \text{if } x_{k*} < x_{ik} < x_k^* \\ 5.5 & \text{if } x_{ik} \leq x_{k*} \end{cases} \tag{3.3}$$

where x_{k*} and x_k^* are the least and most preferred values of criterion k, respectively. With this normalization, the partial evaluation of the banks on a criterion k ranges in a scale from 0.5 (best performance) to 5.5 (worst performance), and the final evaluation model is just a modified version of the net flow model (3.2):

$$V(\mathbf{x}_i) = \sum_{k=1}^{K} w_k v_k(x_{ik}) \in [0.5, 5.5] \tag{3.4}$$

This model can be used to rank the banks in terms of their relative performance, thus providing insight into the strengths and weaknesses of each bank within the competitive market and the conditions that prevail. Given the overall score defined in this way, the associated relative rating is specified by defining the intervals [0.5, 1.5] for group 1, (1.5, 2.5] for group 2, (2.5, 3.5] for group 3, (3.5, 4.5] for group 4 and (4.5, 5.5] for group 5.

It should be noted, however, that while the net flow model (3.2) is purely relational (e.g., the evaluation of a bank is expressed solely in terms of the other banks in the sample), with the introduction of the transformation (3.3), the final evaluation model (3.4) incorporates both relational and absolute aspects. This is because the least and most preferred values of the criteria are not defined on the basis of the banks under consideration. Instead, they represent reference points corresponding to high and low risk bank profiles, defined on the basis of the risk analyst's attitude towards risk. In that respect, as the banking sector is improving, the differences $\phi_k(x_{ik}) - \phi_k(x_k^*)$ will decrease, thus leading to improved ratings. Similarly, as the sector deteriorates as a whole, the differences $\phi_k(x_k^*) - \phi_k(x_{ik})$ will increase, resulting in a deterioration of the ratings. Therefore, the rating score of a bank combines its relative performance as opposed to other banks, as well as the performance of the banking sector as a whole compared to predefined risk profiles. The relative evaluation enables the consideration of the interrelationships and interactions between the banks, which is related to systematic risk.

3.3.2 Absolute Evaluation

Except for the above "hybrid" evaluation process, which combines both relative and absolute elements, a purely absolute evaluation approach can also be realized within the context of the PROMETHEE methodology. In this case the results are based only on the comparison of the banks to a pre-specified reference point, whereas the relative performance of the banks is excluded from the analysis.

In cooperation with the analysts in the Bank of Greece, two options were defined for the specification of the reference point. In the first case the banks are compared to the ideal point (ideal bank). This kind of evaluation provides an assessment of the capability of the banks to perform as good as possible. The second option uses an anti-ideal point. Both the anti-ideal and the ideal point ($\mathbf{x}_*$ and $\mathbf{x}^*$, respectively) are defined by the analysts of the Bank of Greece, each consisting of the least and most preferred values of each criterion, i.e. $\mathbf{x}_* = (x_{1*}, x_{2*}, \ldots, x_{K*})$ and $\mathbf{x}^* = (x_1^*, x_2^*, \ldots, x_K^*)$.

In the case where the banks are compared to the ideal point, the partial evaluation function is adjusted as follows:

$$v_k(x_{ik}) = \begin{cases} 5.5 & \text{if } x_{ik} \leq x_{k*} \\ 0.5 + 5\dfrac{\pi_k(x_k^*, x_{ik})}{\pi_k(x_k^*, x_{k*})} & \text{if } x_{ik} > x_{k*} \end{cases}$$

On the other hand, when the anti-ideal point is used, the following partial evaluation function is used:

$$v_k(x_{ik}) = \begin{cases} 0.5 + 5\dfrac{\pi_k(x_k^*, x_{*k}) - \pi_k(x_{ik}, x_{*k})}{\pi_k(x_k^*, x_{*k})} & \text{if } x_{ik} < x_k^* \\ 0.5 & \text{if } x_{ik} \geq x_k^* \end{cases}$$

3.3.3 Analytic Sensitivity Analysis

Naturally, the multicriteria evaluations defined above incorporate some uncertainty and subjectivity, mainly with regard to the parameters of the PROMETHEE method, which include the criteria weights and the parameters σ_k and p_k of the partial preference functions. Furthermore, since banks operate in a dynamic environment, it is also important to identify changes in the input data that may lead to changes in the rating result. This analysis is performed both for the complete set of banks, as well as for each individual bank separately.

In a first stage, these issues can be addressed by analytic sensitivity procedures. For the criteria weights, the objective of the analysis is to define a range of values for the weight of each criterion k for which the rating of the banks remains unchanged. This can be easily done by imposing the condition that the global score $V(\mathbf{x}_i)$ of each bank i should remain within the score range associated with its rating, as defined with the pre-specified weights.

A similar process can also be employed for the parameters of the criteria preference functions. However, with the pairwise relative evaluation scheme of the PROMETHEE method, the partial preference indices are generally non-monotone and non-convex functions of the corresponding parameters σ and p. Thus, in this case it is not possible to define specific bounds for these parameters within which the rating of the banks does not change. On the other hand, the bounds can be explicitly defined for the absolute evaluation process. In particular, let us assume a bank i which is assigned to the rating group ℓ, defined by a range of scores $(\alpha_\ell, \beta_\ell]$ and suppose that a range $[l_k, u_k]$ should be defined for the parameters of the preference function of a criterion k, such that the rating group of the bank does not change, i.e. $\alpha_\ell < V(\mathbf{x}_i) \leq \beta_\ell$. Then:

$$V(\mathbf{x}_i) > \alpha_\ell \Leftrightarrow v_k(x_{ik}) > \max\left\{0.5, \frac{\alpha_\ell - \sum_{j\neq k} w_j v_j(x_{ij})}{w_k}\right\} \tag{3.5}$$

For illustrative purposes, it can be assumed that: (1) the Gaussian preference function is used, (2) the absolute evaluation is performed in comparison to the ideal point, and (3) $x_{k*} < x_{ik} < x_k^*$. Then, taking into account that $v_k(x_{ik})$ decreases with the preference parameter, and denoting by z_{ik} the left-hand side of (3.5), the upper bound u_k is defined as follows:

$$\begin{aligned}
0.5 + 5\frac{\pi_k(x_k^*, x_{ik})}{\pi_k(x_k^*, x_{k*})} &> z_{ik} \Rightarrow \\
\pi_k(x_k^*, x_{ik}) &> \frac{(z_{ik} - 0.5)\pi_k(x_k^*, x_{k*})}{5} \Rightarrow \\
1 - \exp\left[-\frac{(x_k^* - x_{ik})^2}{2u_k^2}\right] &> \frac{(z_{ik} - 0.5)\pi_k(x_k^*, x_{k*})}{5} \Rightarrow \\
u_k &< \sqrt{\frac{-(x_k^* - x_{ik})^2}{2\ln[1 - 0.2(z_{ik} - 0.5)\pi_k(x_k^*, x_{k*})]}}
\end{aligned}$$

Note that if $z_{ik} > 0.5 + 5/\pi_k(x_k^*, x_{k*})$, then $u_k = +\infty$. The same process is used to define the lower bound l_k:

$$\begin{aligned}
V(\mathbf{x}_i) \leq \beta_\ell \Leftrightarrow v_k(x_{ik}) &\leq \min\left\{5.5, \frac{\beta_\ell - \sum_{j\neq k} w_j v_j(x_{ij})}{w_k}\right\} = o_{ik} \Rightarrow \\
1 - \exp\left[-\frac{(x_k^* - x_{ik})^2}{2l_k^2}\right] &\leq \frac{(o_{ik} - 0.5)\pi_k(x_k^*, x_{k*})}{5} \Rightarrow \\
l_k &\geq \sqrt{\frac{-(x_k^* - x_{ik})^2}{2\ln[1 - 0.2(o_{ik} - 0.5)\pi_k(x_k^*, x_{k*})]}}
\end{aligned}$$

With $l_k = 0$ whenever $o_{ik} < 0.5$.

A similar procedure can also be applied with the linear preference function and the comparison to the anti-ideal point. In addition to the specification of bounds for the parameters of the preference functions, additional information can be obtained by observing the general impact of the preference parameters to the overall evaluation of the banks (as a whole and individually). This is done with the calculation of a sensitivity index Δ_k, which measures the mean maximum percentage change in the global evaluation of the banks due to a change in the preference parameter of criterion k. In particular, let $v_k(x_{ik}, a_k)$ denote the partial performance of bank i on criterion k, expressed as a function of x_{ik} and the criterion's preference parameter a_k. Then, two optimization problems are solved to find the parameter value a_{*ik} (a^*_{ik}) that minimize (maximize), the partial performance of bank i on criterion k, i.e.:

$$v_k^{\min}(x_{ik}, a_{*ik}) = \min_{a_{ik}>0} v_k(x_{ik}, a_{ik}) \quad \text{and} \quad v_k^{\max}(x_{ik}, a^*_{ik}) = \max_{a_{ik}>0} v_k(x_{ik}, a_{ik})$$

Then, the sensitivity index δ_{ik} measuring the impact of criterion's k preference parameter on the global performance of bank i is defined as follows:

$$\delta_{ik} = \max\left\{ w_k \frac{v_k^{\max}(x_{ik}, a^*_{ik}) - v_k(x_{ik})}{V(\mathbf{x}_i)},\ w_k \frac{v_k(x_{ik}) - v_k^{\min}(x_{ik}, a_{*ik})}{V(\mathbf{x}_i)} \right\} \tag{3.6}$$

where $V(\mathbf{x}_i)$ is the global performance of the bank obtained with criterion's k preference parameter defined by the decision-maker and $v_k(x_{ik})$ the corresponding partial score. For instance, a sensitivity index $\delta_{ik} = 0.3$ indicates that a change in the preference parameter of criterion k, may lead to a change of up to 30 % in the global performance of bank i. The direction of the change (decrease or increase) can be easily found by identifying which of the two arguments provides the maximum in (3.6).

The sensitivity index Δ_k is then calculated as:

$$\Delta_k = \frac{1}{M} \sum_{i=1}^{M} \delta_{ik}$$

In the case of absolute evaluation $v_k^{\min}(x_{ik}, a_{*ik})$ and $v_k^{\max}(x_{ik}, a^*_{ik})$ are easy to find because $v_k(x_{ik}, a_k)$ is a monotone function of a_k, and the extremes are found by imposing a range of reasonable values for a_k (e.g., between 0.001 and 100). On the other hand, in the relative evaluation process, $v_k(x_{ik}, a_k)$ is generally a non-convex function of a_k. In this case, a simple genetic algorithm is employed in order to find $v_k^{\min}(x_{ik}, a_{*ik})$ and $v_k^{\max}(x_{ik}, a^*_{ik})$.

3.3.4 Robustness Analysis Through Simulation

The analytic procedures described in the previous section, provide useful local information about the sensitivity of the rating results. Further information can be derived through simulation approaches to obtain a holistic view of the robustness of the results. Simulation is used to analyze the robustness of the ratings with respect to changes in the weights of the criteria, but the process can be easily extended to consider the parameters of the preference functions, too.

The simulation involves the generation of multiple scenarios regarding the weights of the criteria. Two options can be considered for the generation of the weights. In the first case, the weights are generated at random over the unit simplex. Alternatively, the decision maker can provide a ranking of the criteria according to their relative importance, and then random weights are generated, which are in accordance with the ordering of the criteria.

The results of the simulation can be analyzed in terms of the mean and median of the global performance scores, their standard deviation and confidence intervals. Furthermore, for each individual bank useful conclusions can be drawn on the distribution of its rating under different weighting scenarios.

3.3.5 Implementation

The proposed multicriteria methodology has been implemented in an integrated decision support system (DSS) [73]. The system enables multiple users (senior or junior level analysts) to work simultaneously on a common data base. Senior bank analysts are responsible for defining the evaluation criteria and setting the main parameters of the evaluation process (criteria weights, the type of the criteria preference functions, and preference parameters). Lower level analysts have full access to all features of the multicriteria evaluation process, but they are not allowed to perform permanent changes in the evaluation parameters.

Except for data base management and the use of the multicriteria tools, the DSS includes a user-friendly interface that facilitates the preparation of several reports in graphical and tabular format. The system also includes multivariate statistical analysis techniques such as principal components analysis (PCA) as well as some additional modules that support analysts in the specification of the criteria weights using the rank-order centroid (ROCD) and rank-sum (RS) approaches [131]. PCA is a multivariate statistical analysis tool that enables the examination of the explanatory power of the criteria from a statistical point of view. On the other hand, the ROCD and RS approaches simplify the definition of the weights of the criteria. Both techniques only require the user to define a weak-order of the criteria according to their relative importance, without asking for the specification of the exact trade-offs. The ROCD estimates are derived by the centroid of the polyhedron defined by the constraints on the criteria trade-offs, whereas the RS approach relies on the order statistics of the uniform distribution. In particular, assuming that K criteria have been ranked from the most to the least important ones (criterion 1 is assumed to be the most important and criterion K the least important one), the ROCD and RS weights for criterion k are defined as follows:

$$\text{ROCD weight: } w_k = \frac{1}{K}\sum_{\ell=k}^{K}\frac{1}{\ell} \qquad \text{RS weight: } w_k = \frac{K+1-k}{0.5K(K+1)}$$

The system runs on any MS Windows-based PC and it is currently used by the Risk Analysis and Supervisory Techniques Division of the Bank of Greece for evaluating and monitoring the strengths and weaknesses of Greek banks, on the basis of the supervisory policy defined in accordance with the international regulatory framework.

The next section presents an illustrative application of the methodology on sample data for Greek commercial banks over the period 2001–2005.

3.4 Application

3.4.1 Data and Evaluation Parameters

The data involve detailed information for all Greek banks during the period 2001-2005. Overall, 18 banks are considered. The banks are evaluated on a set of 31 criteria (Table 3.1), selected in co-operation with expert analysts of the Bank of Greece, who are responsible for monitoring and evaluating the performance of the banks. The criteria are organized into six categories (capital, assets, management, earnings, liquidity, sensitivity to market risks), in accordance with the CAMELS framework. Overall, 17 quantitative and 14 qualitative criteria are used. By "quantitative"/"qualitative" criteria, we refer to criteria used to evaluate the financial and non-financial, respectively, aspects of the operation of banks. All criteria are actually measured in numerical scales. For the qualitative criteria an interval 0.5–5.5 scale is used (with lower values indicating higher performance), in accordance with the existing practice followed by the risk analysts of the Bank of Greece, who are responsible for collecting and evaluating the corresponding information.

The weights of each category of criteria and the criteria therein have been defined by the expert analysts of the Bank of Greece. Table 3.2 presents the weights defined for each category of criteria along with the corresponding ROCD and RS estimates defined using the ordering of the criteria according to the expert's weights. It is interesting to note that the RS estimates are very close to the actual relative importance of each criteria group. The same was also observed at the individual criteria level. Overall, the quantitative criteria are assigned a weight of 70 %, with the remaining 30 % involving qualitative criteria.

Table 3.1 Evaluation criteria

Categories	Abbr.	Criteria
Capital	Cap1	Capital adequacy ratio
	Cap2	TIER II capital/TIER I
	Cap3	Qualitative
Assets	Ass1	Risk-weighted assets/assets
	Ass2	Non performing loans – provisions/Loans
	Ass3	Large exposures / (TIER I + TIER II capital)
	Ass4	[0.5(Non performing loans) – provisions]/equity
	Ass5	Qualitative
Management	Man1	Operating expenses/operating income
	Man2	Staff cost/assets
	Man3	Operating income/business units
	Man4	Top management competencies, qualifications and continuity
	Man5	Managers' experience and competence
	Man6	Resilience to change, strategy, long term horizon
	Man7	Management of information systems
	Man8	Internal control systems
	Man9	Financial risk management system
	Man10	Internal processes charter – implementation monitoring
	Man11	Timely and accurate data collection
	Man12	Information technology systems
Earnings	Ear1	Net income/assets
	Ear2	Net income/equity
	Ear3	Interest revenue/assets
	Ear4	Other operating revenue/assets
	Ear5	Qualitative
Liquidity	Liq1	Cash/assets
	Liq2	Loans – provisions/deposits
	Liq3	Real funding from credit institutions/assets
	Liq4	Qualitative
Market	Mar1	Risk-weighted assets II/Risk-weighted assets (I and II)
	Mar2	Qualitative

Table 3.2 Weights of each category of criteria

Categories	Weight	ROCD weights	RS weights
Capital	30	47.92	30.77
Assets	20	22.92	23.08
Management	15	10.42	15.38
Earnings	15	10.42	15.38
Liquidity	10	4.17	7.69
Market	10	4.17	7.69

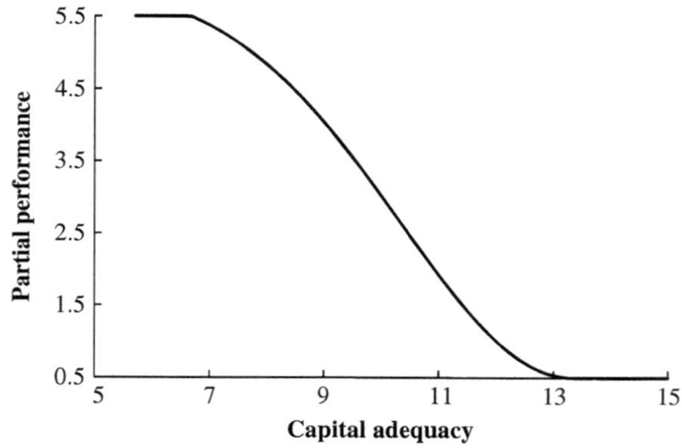

Fig. 3.2 The partial performance function for the capital adequacy ratio (absolute evaluation)

All quantitative criteria are considered through the Gaussian preference function, whereas a linear preference function is used for the qualitative criteria. Figure 3.2 illustrates the partial performance function for the capital adequacy ratio. The function decreases with the values of the criterion, thus indicating that higher capital adequacy values are associated with higher performance and lower risk. The least and most preferred values have been set by the expert analysts to 6.67 and 13.33, respectively. Thus, banks with capital adequacy ratio higher than 13.33 achieve a partial score of 0.5, whereas high risk banks with capital adequacy ratio below 6.67 have a partial score of 5.5. In all cases, the preference parameters have been set in such a way so as to ensure that the partial scores of the banks span, as much as possible, the whole range of values in the pre-specified score range [0.5, 5.5].

3.4.2 Results

Table 3.3 presents the overall evaluation results using the relative assessment procedure. Similar results are also obtained with the absolute evaluation process.[1] The results indicate that most banks achieved a rating grade of 2 or 3, each corresponding to performance scores in (1.5, 2.5] and (2.5, 3.5], respectively. There is no bank in the first (best) grade (score ≤ 1.5) nor in the highest (5th) risk grade (scores >4.5).

The dynamics of the performance scores of the banks, indicate that no significant changes are observed between the 5 years of the analysis. Nevertheless, 2002 appears to have been the worst year; compared to 2001 only two banks managed to improve their performance. In 2003 most banks improved their performance (compared to 2002). In 2004 and 2005 no noticeable trend is observed. The highest performance

[1] On average, the rating scores with the absolute evaluation using the ideal reference point were lower (better) compared to the relative evaluation (average difference −0.064). Throughout the 5 years, the ratings were identical in 92 % of the cases with 2 downgrades and 5 upgrades. Similarly, the rating scores with the absolute evaluation using the anti-ideal reference point, were on average higher (worse) compared to the relative evaluation (average difference 0.06). Throughout the 5 years, the ratings were identical in 87 % of the cases with 13 downgrades and none upgrade.

Table 3.3 Overall evaluation results (relative assessment)

Banks	2001	2002	2003	2004	2005
A1	2.59	3.38	2.82	2.65	2.33
A2	2.43	2.48	1.89	1.77	2.03
A3	2.70	3.26	3.35	3.21	3.04
A4	3.19	3.01	2.71	2.91	3.15
A5	2.29	2.45	2.43	2.54	2.52
A6	2.09	2.88	3.02	3.07	3.04
A7	2.03	2.18	1.70	1.63	1.61
A8	1.60	2.00	2.09	1.93	1.95
A9	–	2.10	2.31	2.32	2.59
A10	2.85	3.67	3.42	3.74	3.43
A11	2.31	2.82	2.52	2.17	2.52
A12	2.34	2.49	2.35	2.39	2.36
A13	2.16	2.20	2.26	2.75	2.13
A14	–	2.28	2.18	2.62	3.78
A15	–	2.58	2.40	2.44	2.50
A16	2.64	2.58	2.40	2.27	2.24
A17	–	2.18	2.40	1.98	2.03
A18	–	2.49	2.24	2.32	1.95

improvements have been achieved by banks A7 (20.7 % improvement in 2005 compared to 2001) and A18 (21.4 % improvement in 2005 compared to 2002). On the other hand, the highest decreases in performance involve banks A14 and A6. Bank A14 is the only bank that has been downgraded by more than one rating point during the examined time period. In 2002 (the first year being evaluated) bank A14 was assigned in the 2nd risk grade, deteriorated in the 3rd grade in 2004 and then in the 4th grade in 2005. This downgrade has been mainly due to the deterioration of the assets quality and the weakening of the earnings of the bank.

Table 3.4 provides some sensitivity analysis results for each category of criteria. The presented results involve the weights ranges within which the rating of the banks remains unchanged in each year. When compared to the pre-specified weights of each category of criteria, it becomes apparent that the rating of the banks is most sensitive to changes in the relative importance of the capital dimension. The earnings dimension also seems to be critical (mainly in 2002 and 2003). On other hand, the relative importance of the management dimension is the least likely to alter the rating of the banks. Overall, the ratings in 2002 and 2005 seem to be the most sensitive to changes in the relative importance of the criteria categories, since the obtained bounds are generally closer to the pre-specified weights. As far as the individual criteria are concerned, the most critical ones (as far as their weighting is concerned) were found to be Cap1 (capital adequacy ratio) and Mar1 (risk-weighted assets II / risk-weighted assets I and II). The same two criteria were also found to have among the highest sensitivity indices, particularly in the most recent years (2004–2005). In general, the sensitivity indices were found to be limited (lower than 4 % in all cases).

Table 3.4 Sensitivity analysis results

Categories	Weight	2001	2002	2003	2004	2005
Capital	30	[21.9, 36.5]	[29.4, 31.8]	[25.3, 33.9]	[25.4, 34.8]	[29.9, 32]
Asset	20	[11.7, 29.1]	[17.8, 23]	[4.2, 24.5]	[13.4, 34.2]	[0, 20.5]
Management	15	[0.3, 29.6]	[12.2, 16]	[0.0, 23.1]	[0.9, 22.7]	[12.3, 15.4]
Earnings	15	[7.2, 23.4]	[11.3, 15.7]	[13.4, 20.1]	[5.9, 21.2]	[13.7, 15.2]
Liquidity	10	[4.2, 22.2]	[4.3, 11.6]	[8.9, 14.1]	[6.4, 14.4]	[8.4, 10.1]
Market risk	10	[0, 18.9]	[8.3, 10.9]	[5.3, 11.9]	[4.2, 13.1]	[9.8, 11.4]

On the other hand, in the case of absolute evaluation the impact of the preference parameters was higher, with sensitivity indices up to 8.5 %.

Further results on the sensitivity of the ratings to the weighting of the criteria are obtained with Monte Carlo simulation. The simulation is based on 1,000 different weighting scenarios. In each simulation run, a weighting vector is generated at random, but taking into account the ranking of the criteria according to their importance as defined by the expert analysts. Summary results for 2005 are presented in Table 3.5. The results involve statistics on the global performance score of the banks (mean, median 95 % confidence interval), as well as the distribution of the ratings for each bank. The obtained results are in accordance with the ones given earlier in Table 3.3. In most cases, the rating of the banks is quite robust under different weighting scenarios. The most ambiguous cases involve banks A5, A9, A10, A11 and A15. Future revisions of the rating process or changes in the input data for these banks are highly likely to affect their ratings.

Banks A10 and A14 are the only ones for which a high risk rating seems quite applicable. Further analysis for these two high risk banks is performed by examining the correlations between the randomly generated criteria weights and the global performance of the banks, throughout the simulation experiment. Table 3.6 summarizes the results for the most influential criteria, i.e., the ones whose weight has the highest absolute correlation with the performance of the banks. Criteria with negative correlations are associated with the strong points of the banks, in the sense that an increase in the weight of these criteria leads to a decrease in the global performance score of the banks, thus to lower (better) rating. On the other hand, criteria with positive correlations indicate the weaknesses of the banks, in the sense that an increase in the weight of these criteria leads to an increase in the global performance score of the banks, thus to higher (worse) rating. The obtained results show that the major weaknesses of bank A10 involve its exposure to liquidity risk and its weak earnings. On the other hand, its exposure to market risk is limited, thus leading to an improvement of its overall performance. The exposure to market risk is also a strength for bank A14, which seems to suffer from poor earnings, low asset quality and low capital adequacy.

Table 3.5 Simulation results for 2005

	Statistics				Rating distribution				
Banks	Mean	Median	95 % CI		1	2	3	4	5
A1	2.36	2.37	2.05	2.62	0.0	83.2	16.8	0.0	0.0
A2	2.02	2.03	1.57	2.39	0.7	99.1	0.2	0.0	0.0
A3	3.11	3.10	2.85	3.37	0.0	0.0	99.8	0.2	0.0
A4	3.17	3.17	2.86	3.45	0.0	0.0	98.6	1.4	0.0
A5	2.55	2.56	2.26	2.80	0.0	34.8	65.2	0.0	0.0
A6	3.00	3.00	2.73	3.29	0.0	0.0	100.0	0.0	0.0
A7	1.68	1.69	1.32	2.00	20.2	79.8	0.0	0.0	0.0
A8	1.91	1.92	1.48	2.29	2.9	97.1	0.0	0.0	0.0
A9	2.55	2.56	2.23	2.84	0.0	35.1	64.9	0.0	0.0
A10	3.48	3.48	3.17	3.78	0.0	0.0	56.1	43.9	0.0
A11	2.48	2.48	2.21	2.73	0.0	55.5	44.5	0.0	0.0
A12	2.38	2.37	2.15	2.64	0.0	82.1	17.9	0.0	0.0
A13	2.08	2.08	1.77	2.38	0.1	99.7	0.2	0.0	0.0
A14	3.75	3.74	3.39	4.16	0.0	0.0	11.0	89.0	0.0
A15	2.52	2.53	2.13	2.85	0.0	43.7	56.3	0.0	0.0
A16	2.18	2.18	1.93	2.42	0.0	99.6	0.4	0.0	0.0
A17	2.01	2.01	1.77	2.24	0.0	100.0	0.0	0.0	0.0
A18	1.91	1.91	1.54	2.27	1.6	98.4	0.0	0.0	0.0

Table 3.6 Correlations between the criteria weights and the performance of banks A10, A14

A10		A14	
Mar1	−52.2	Cap2	−56.6
Ass4	−40.4	Mar1	−48.1
Cap1	−33.0	Mar2	−26.6
Mar2	−26.9	Liq3	−13.4
Liq2	20.7	Ass3	−11.5
Liq3	24.6	Ear1	18.7
Ass2	32.1	Ear2	22.7
Ear1	32.2	Ass2	22.8
Liq1	32.6	Ass4	24.0
Ear2	34.5	Cap1	54.7

3.5 Bank Efficiency Versus Bank Performance

The analysis of efficiency in the banking sector has been a major research topic in the area of banking management. According to production theory, efficiency is defined as the ratio between the outputs of a production unit over the inputs used in the production process.

In the context of banking management, the efficiency of banks can be considered under a profit or an intermediation approach [77, 195]. The profit approach focuses on the ability of a bank to control its costs in order to maximize its profits. In this setting, revenue components are the outputs and cost components define the inputs. The intermediation approach, on the other hand, focuses on the ability of a bank to produce financial services (e.g., providing loans to customers, the investment activities of the bank, etc.) using its available resources (personnel, fixed assets, loans, deposits, equity, etc.).

Efficiency assessments are based on frontier methods, with non-parametric techniques such as data envelopment analysis (DEA) being widely used [88]. DEA provides estimates of the relative efficiency for a set of decision making units (i.e., banks), based on their inputs and outputs.

In particular, let $\mathbf{X}$ be a $K \times M$ data matrix for K input variables of M DMUs and $\mathbf{Y}$ be a $O \times M$ matrix for O outputs. Then, the efficiency of the ith DMU is measured by the ratio:

$$\theta_i = \frac{\mathbf{u}_i \mathbf{y}_i}{\mathbf{y}_i \mathbf{x}_i} \in [0, 1]$$

where $\mathbf{x}_i$ and $\mathbf{y}_i$ are, respectively, the available data for the inputs and outputs of DMU i, whereas $\mathbf{u}_i, \mathbf{v}_i \geq \mathbf{0}$ are weight vectors corresponding to these input/outputs.

DEA provides an assessment of the relative efficiency of a DMU compared to a set of other DMUs. In this relative evaluation setting, each DMU is free to specify its own combination of input-output weights that maximize its performance relative to its peers (i.e., competitors). Under constant returns to scale (CRS) and assuming an input orientation, the optimal efficiency for the ith DMU can be estimated through any of the two following linear programming formulations (CCR model, [43]):

$$\begin{array}{ll}
Primal: & Dual: \\
\max \mathbf{u}_i \mathbf{y}_i & \min \theta_i^C \\
\text{s.t. } \mathbf{v}_i \mathbf{X} - \mathbf{u}_i \mathbf{Y} \geq \mathbf{0} & \text{s.t. } \theta_i^C \mathbf{x}_i - \mathbf{X}\boldsymbol{\lambda} \geq \mathbf{0} \\
\quad \mathbf{v}_i \mathbf{x}_i = 1 & \quad \mathbf{Y}\boldsymbol{\lambda} \geq \mathbf{y}_i \\
\quad \mathbf{u}_i, \mathbf{v}_i \geq \mathbf{0} & \quad \boldsymbol{\lambda} \geq \mathbf{0}, \ \ \theta_i^C \in \mathbb{R}
\end{array} \tag{3.7}$$

The estimate θ_i^C obtained from the CCR model provides a global technical efficiency measure without taking into consideration any scale effects. In that sense, it is assumed that all DMUs are operating at an optimal scale [46]. To take into account cases where this assumption is not true, variable returns to scale (VRS) can be introduced by simply adding the convexity constraint $\lambda_1 + \lambda_2 + \cdots + \lambda_N = 1$ to the dual CCR model. This constraint ensures that a DMU is benchmarked only against other units of similar size. The resulting model is known as the BCC model [17].

The combination of the results obtained from the CCR and BCC models provides a decomposition of the global efficiency as follows:

$$\theta_i^C = \theta_i^V \theta_i^S$$

where $0 \leq \theta_i^V \leq 1$ is the pure efficiency score obtained under VRS from the BCC model and $0 \leq \theta_i^S \leq 1$ is the scale efficiency factor. Thus, the inefficiency of a DMU can be attributed to inefficient operation (e.g., too small θ_i^V), disadvantageous exogenous conditions (corresponding to scale inefficiency), or both.

The methodological framework of efficiency analysis with DEA has significant similarities but also notable differences with the framework of MCDA. For instance, Joro et al. [136] focused on the connections between DEA and multiobjective optimization, and noted that both fields are interested in identifying efficient points and projecting inefficient units to the efficiency frontier. However, in DEA the projections are made with "optimally" selected weights which differ for each DMU, whereas in MCDA predefined preferential weights are used for all cases under consideration. In that regard, the authors considered multicriteria methods as *ex ante* planning tools and DEA as an *ex post* evaluation tool.

Furthermore, several authors have suggested using DEA for multicriteria evaluation purposes, given that DEA is a data-driven approach that requires minimal information [214]. However, most of such DEA-based evaluation models (e.g., cross-efficiency and super-efficiency models) have significant methodological shortcomings [32].

In the area of banking management, DEA models are useful for evaluating the relative efficiency of banks and discriminating between efficient and inefficient banks. Nevertheless, efficiency is only one aspect of the overall performance and risk of the banks. For instance, one cannot assume that all efficient banks are performing equally well. The same applies to inefficient banks as direct comparisons among such cases are generally meaningful only for those sharing similar characteristics (i.e., belonging to the same facet of the efficient frontier). Furthermore, important aspects, such as capital adequacy, risk management systems, liquidity, and external conditions, are only indirectly relevant to efficiency assessments, whereas in a performance evaluation setting they are considered as key issues. Finally, a bank performance assessment model should be transparent and allow the analysis of all banks in a common setting. In DEA, however, the assessment model does not have a well-defined functional form. Instead, it is based on the solution of a linear program, the analysis is meaningful only for samples of adequate size (it is pointless to perform multidimensional efficiency comparisons for sample with only a few banks), and the weightings of the input/output variables differ for each bank. Even though these properties make sense in an efficiency analysis setting, from an performance evaluation point of view they are troublesome.

MCDA models, on the other hand, are appropriate for benchmarking purposes, allowing the consideration of all pertinent factors that describe (direct or indirectly) the performance of banks, and enabling comparisons to be performed over time based on well-defined functional, relational, or symbolic models, that a bank analyst can use in a straightforward manner.

With the above remarks in mind, it is meaningful to explore the synergies between efficiency analysis based on frontier methods and performance assessment models constructed with MCDA techniques.

Chapter 4
Credit Scoring

Abstract The recent financial crisis has highlighted the importance of credit risk assessment for financial institutions, firms, and supervisors. Credit scoring systems are important tools for credit risk evaluation and monitoring. This chapter describes the process for building and testing credit scoring models and illustrates how multicriteria techniques based on disaggregation analysis can be used in this area. Empirical results are also presented, derived from an application to a large sample of Greek firms.

Keywords Credit risk · Preference disaggregation · Multicriteria classification

4.1 Credit Scoring Systems

Credit risk modeling plays a crucial role in financial risk management, in areas such as banking, corporate finance, and investments. Credit risk management has evolved rapidly over the past decades, but the global credit crisis of 2007–2008 highlighted that there is still much to be done at multiple levels. Altman and Saunders [3] list five main factors that have contributed to the increasing importance of credit risk management:

1. the worldwide increase in the number of bankruptcies,
2. the trend towards disintermediation by the highest quality and largest borrowers,
3. the increased competition among credit institutions,
4. the declining value of real assets and collateral in many markets, and
5. the growth of new financial instruments with inherent default risk exposure, such as credit derivatives.

Credit risk refers to the probability that an obligor will not be able to meet scheduled debt obligations (i.e., default). Early credit risk management was primarily based on empirical evaluation systems of the creditworthiness of a client. CAMEL has been

M. Doumpos and C. Zopounidis, *Multicriteria Analysis in Finance*,
SpringerBriefs in Operations Research, DOI: 10.1007/978-3-319-05864-1_4,

the most widely used system in this context, which is based on the empirical combination of several factors related to capital, assets, management, earnings, and liquidity. It was soon realized however, that such empirical systems cannot provide a solid and objective basis for credit risk management. This led to an outgrowth of studies from academics and practitioners on the development of new credit risk assessment systems. These efforts were also motivated by the changing regulatory framework that now requires banks to implement specific methodologies for managing and monitoring their credit portfolios [18].

The existing practices are based on sophisticated analytic modeling techniques, which are used to develop a complete framework for measuring and monitoring credit risk. Credit scoring systems are in the core of this framework and are widely used to assess the creditworthiness of firms and individuals, estimate the probabilities of default, and classify the obligors into risk groups.

The aim of credit scoring models is to assess the probability of default for an obligor and differentiate individual credits by the risk they pose. This allows creditors to monitor changes and trends in risk levels, thus promoting safety and soundness in the credit granting process. Credit scoring models are also used for credit approval and underwriting, loan pricing, relationship management and credit administration, allowance for loan and lease losses and capital adequacy, credit portfolio management and reporting [48].

Generally, a credit scoring model can be considered as a mapping function $F(\mathbf{x}; \boldsymbol{\alpha})$, defined by a vector of modeling parameters $\boldsymbol{\alpha}$, such that $F(\mathbf{x}; \boldsymbol{\alpha}): \mathbb{R}^K \to G$. The credit scoring model provides estimates for the probability of default for an obligor described by a vector $\mathbf{x} \in \mathbb{R}^K$ of K attributes and maps the result to a set G of risk categories.

The attribute vector $\mathbf{x}$ represents all the relevant information that describes the obligor, including financial and non-financial data. For instance, for corporate loans, financial ratios, measuring the company's profitability, liquidity, leverage, etc., are usually considered to be important quantitative attributes. Non-financial criteria are related to the company's activities, its market position, management quality, growth perspectives, credit history, the trends in its business sector, etc. Empirical evidence has shown that such non-financial attributes significantly improve the estimates of credit scoring and default prediction models [107]. Furthermore, market data and estimates from the Black-Scholes-Merton model have also been shown to be strong predictors of credit risk [68, 250].

Credit risk assessments can be obtained either through models developed internally by financial institutions [244] or are provided externally by credit rating agencies (CRAs). The latter, provide credit ratings for firms in a multi-grade risk scale. Despite the criticisms on their scope and accuracy [93, 190, 241], they are widely used by investors, financial institutions, and regulators, and they have been extensively studied in academic research [130]. However, external ratings, even if considered to be reliable, they do not have a global coverage as they are available only for large corporations, they are not always provided in a timely manner, and they do not differentiate between companies in the same rating class. On the other hand, credit scoring models provide a unique credit score to each rated borrower and they are applicable

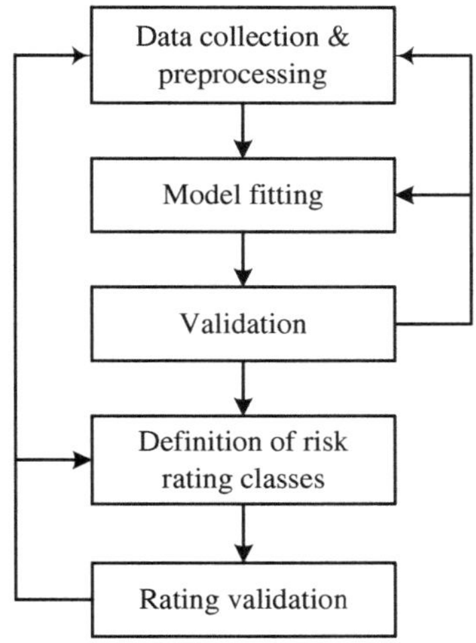

Fig. 4.1 The process for developing credit rating models

to all borrowers (including corporate loans and consumer credit), thus providing a full coverage of a loan portfolio.

4.2 Construction and Validation Process

The credit scoring modeling process can be described through the five steps illustrated in Fig. 4.1.

The process begins with the collection of appropriate data involving obligors with known creditworthiness status. In a typical setting, data for defaulted and non-default cases are collected. These data can be obtained from the historical data base of a credit institution or from external sources. At this stage, some preprocessing of the data is necessary in order to transform them into meaningful attributes, to eliminate outliers, and to select the appropriate set of attributes for the analysis. These steps lead to the final data $\{\mathbf{x}_i, y_i\}_{i=1}^m$, where $\mathbf{x}_i$ is the input attribute vector for obligor i, y_i in the known status of the obligor, and m in the number of observations in the data set. These data, which are used for model development, are usually referred to as *training data*.

The second stage involves the model fitting process, which refers to the identification of the model's parameters that best describe the training data. For instance, assume the following linear model:

$$F(\mathbf{x}) = \alpha_0 + \mathbf{x}\boldsymbol{\alpha}$$

where $\boldsymbol{\alpha} \in \mathbb{R}^K$ is the vector with the coefficients of the selected attributes and α_0 is a constant term. In this case, model fitting is involved with finding the optimal parameters $\boldsymbol{\alpha}$ and α_0 on the basis of the information provided by the training data.

This can be expressed as an optimization problem of the following general form:

$$\min_{\alpha \in \mathscr{A}} L(\alpha, \mathbf{X}) \tag{4.1}$$

where $\mathscr{A}$ is a set of constraints that define the feasible (acceptable) values for the parameters of the model, $\mathbf{X}$ is the training data set and L is a loss function measuring the differences between the model's output and the given classification of the training observations.

On the algorithmic side, several statistical, data mining, and operations research techniques are used to implement the model fitting process. The most widely used methods include logistic regression and probit models, but non-parametric techniques have also gained much interest among researchers and practitioners. Some examples include, neural networks, rule-induction algorithms, support vector machines, fuzzy models, ensembles, and hybrid systems (e.g., neuro-fuzzy models). Comprehensive reviews and discussion of popular methods can be used in Abdou and Pointon [1], Crook et al. [54], and Papageorgiou et al. [193].

The result of the model optimization process are validated using another sample of obligors with known status. This is referred to as the *validation sample*. Typically it consists of cases different than the ones of the training sample and for a future time period. The optimal model is applied to these new observations and its predictive ability is measured, usually using statistical measures (for an overview see [228]). The economic aspects of the model's predictive results are also important [30, 92, 147, 188].

The validation of the scoring model is followed by mapping the model's outputs (credit scores) to risk rating classes consisting of borrowers with similar levels of creditworthiness [157]. The defined rating needs also to be validated in terms of its stability over time, the distribution of the obligors in the rating groups, and the consistency between the estimated probabilities of default in each group and the empirical ones which are taken from the population of rated obligors.

4.3 Multicriteria Aspects of Credit Scoring

From the methodological point of view, credit scoring for business and consumer loans is a statistical pattern classification problem, as the decision models are constructed on the basis of historical default data.[1] Nevertheless, some features that analysts often require scoring models to have [147], make MCDA techniques appealing in this context. In particular:

- *Credit scoring models are usually required to be monotone with respect to the inputs*. From an economic and business perspective, the monotonicity assumption

[1] In other specialized credit granting contexts (e.g., project finance), the risk assessment process is mostly based on empirical quantitative and qualitative models [181] (Chaps. 8, 10), which fit well the context of MCDA.

implies that as the input information for a given applicant improves, the estimated probability of default should decrease. Assuming that all attributes are in a maximization form, the monotonicity assumption can be formally expressed as follows:

$$\Pr(D|\mathbf{x}_i) \leq \Pr(D|\mathbf{x}_j), \quad \forall \mathbf{x}_i \succ \mathbf{x}_j \tag{4.2}$$

where $\Pr(D|\mathbf{x}_i)$ is the estimated probability of default for credit applicant i and $\succ$ represents the dominance relationship, defined as follows: $\mathbf{x}_i \succ \mathbf{x}_j \Leftrightarrow \mathbf{x}_i \geq \mathbf{x}_j$ and $x_{ik} > x_{jk}$, for at least one attribute k.
Models that violate monotonicity in an arbitrary manner may fail to be accepted, simply because they lack economic sense, thus providing counterintuitive results from an economic perspective. Furthermore, empirical results have shown that introducing monotonicity in credit scoring models actually improves their predictive performance and robustness, through the elimination of the over-fitting effect [72].

- *Credit scoring models should be transparent and comprehensible.* The predictive accuracy of credit scoring models is not the sole decisive factor for their success in practice. In addition to being accurate, the modes should also be easy to understand by analysts, end users, and regulators. A comprehensible model enables its user to understand its underlying logic and provide justifications on its recommendations [170, 172], instead of simply being used as a black-box analytic recommendation tool.
- *Risk grades are ordinal.* This is often ignored by many popular statistical and computational intelligence techniques used for model building, which often assume that the classes are nominal (i.e., in no particular order).

Multicriteria decision models fit well these requirements: (a) they are by definition ordinal, (b) they provide evaluation results that are monotone with respect to the evaluation criteria, and (c) they promote transparency, enabling the credit analyst to calibrate them on the basis of his/her expert domain knowledge, and allow for justification of the obtained results. Among others, MCDA methods have been used in the area of credit scoring (and the relevant field of bankruptcy prediction) in different ways:

1. As tools for building accurate and transparent credit scoring systems, customized to the needs of particular financial institutions [51, 99]. This is particularly important for special types of credit (e.g., project finance) for which historical data may be lacking. In such cases, MCDA methods can greatly enhance peer expert judgment scoring systems, facilitating the structuring of the credit granting evaluation process and providing formal procedures for aggregating multiple credit evaluation criteria.
2. In combination with other modeling and learning techniques, including rough sets, fuzzy models, case-based reasoning, and neural networks [39, 120, 252, 262]. Such computational intelligence techniques provide strong data analysis capabilities. MCDA on the other hand, provides axiomatic decision models of different

forms. The combination of these paradigms [64] provides a new set of powerful hybrid systems for credit scoring.
3. As optimization approaches for model fitting under multiple performance measures [113, 154, 186]. The performance of a credit scoring model has different aspects, including statistical (e.g., different measures of predictive accuracy) and economic (profit/costs derived from actions taken on the basis of the results of a credit scoring model). Multiobjective optimization techniques enable the consideration such multiple performance measures when building a credit scoring model.
4. As alternatives to popular statistical and machine learning approaches providing more accurate rating results [69, 74, 121]. The results from several studies show that credit scoring models constructed using MCDA preference disaggregation techniques provide robust and accurate results, and often actually outperform other popular approaches. Thus, they could be considered as potential candidates for constructing credit scoring and rating models.

The next section illustrates the application of a multicriteria methodology for developing a credit scoring model and its comparison to popular statistical and non-parametric techniques.

4.4 Using Preference Disaggregation Analysis for the Construction of a Credit Scoring Model

MCDA provides a variety of approaches for credit risk modeling and the construction of credit scoring systems, including outranking techniques [74, 121], rule-based models [39, 61, 252, 262], and value models [51, 69, 63].

To facilitate the presentation we shall focus on additive value models in the framework of the UTADIS method [70, 265]. Additive models are popular approaches for credit risk modeling, as they are intuitive scoring systems, that are simple to understand and implement, as they are compatible with the scorecard structure of credit rating systems used in practice [219]. For instance, Krahnen and Weber [147] conducted a survey among major German banks and found that all of them used credit scoring models expressed in the form of an additive value function:

$$V(\mathbf{x}_i) = \sum_{k=1}^{K} w_k v_k(x_{ik}) \tag{4.3}$$

where the global value $V(\mathbf{x}_i)$ is an estimate of the overall creditworthiness and default risk of obligor i.

In this model, the overall assessment is a weighted average of partial scores $v_1(x_{i1}), \ldots, v_K(x_{iK})$ defined over a set of K credit risk assessment criteria. Without loss of generality, we shall assume that the weighting trade-off constants are non-negative and normalized such that $w_1 + w_2 + \cdots + w_K = 1$. On the other hand, the marginal value functions $v_1(\cdot), \ldots, v_K(\cdot)$, which define the partial scores, are scaled such that $v_k(x_{k*}) = 0$ and $v_k(x_k^*) = 1$, where x_{k*} and x_k^* are the most and least risky level of risk attribute k, respectively. For simplicity, henceforth it will be assumed that all risk assessment criteria are expressed in maximization form (thus implying that all marginal value functions are non-decreasing).

The construction of the credit scoring model (4.3) can be simplified by setting $u_k(x_k) = w_k v_k(x_k)$, which leads to a rescaled set of marginal value functions $u_1, \ldots, u_K$ normalized in $[0, w_k]$. With this transformation, the evaluation model (4.3) can be re-written in the following equivalent form:

$$V(\mathbf{x}_i) = \sum_{k=1}^{K} u_k(x_{ik}) \tag{4.4}$$

This decision model can be linear or nonlinear depending on the form of the marginal value functions. The marginal value functions can be either pre-specified by the decision maker or inferred directly from the data using a preference disaggregation approach. In the context of credit scoring the latter approach is the preferred one, particularly when there are historical data available for constructing the model. Under this scheme, a convenient and flexible way to take into consideration a wide class of monotone marginal value functions, is to assume that they are piecewise linear. In that regard, the range of each risk criterion k is split into $s_k + 1$ subintervals defined by s_k break-points $\beta_0^k < \beta_1^k < \cdots < \beta_{s_k+1}^k$, between the least and the most preferred levels of the criterion (denoted by β_0^k and $\beta_{s_k+1}^k$, respectively), as illustrated in Fig. 4.2. Thus, the marginal value of any alternative i on criterion k can be expressed as:

$$u_k(x_{ik}) = \sum_{r=1}^{s_k} p_{ik}^r d_{kr} \tag{4.5}$$

where $d_{kr} = u_k(\beta_r^k) - u_k(\beta_{r-1}^k) \geq 0$ is the difference between the marginal values at two consecutive levels of criterion k and

$$p_{ik}^r = \begin{cases} 0 & \text{if } x_{ik} < \beta_{r-1}^k \\ \frac{x_{ik} - \beta_{r-1}^k}{\beta_r^k - \beta_{r-1}^k} & \text{if } x_{ik} \in [\beta_{r-1}^k, \beta_r^k] \\ 1 & \text{if } x_{ik} > \beta_r^k \end{cases} \tag{4.6}$$

With the above piecewise linear modeling of the marginal value functions, the scoring model (4.4) can be expressed as a linear function of the step differences in the marginal values between consecutive break-points in the criteria's scale:

$$V(\mathbf{x}_i) = \sum_{k=1}^{K} \mathbf{p}_{ik}^\top \mathbf{d}_k \tag{4.7}$$

where $\mathbf{p}_{ik} = (p_{ik}^1, p_{ik}^2, \ldots, p_{ik}^{s_k})$ and $\mathbf{d}_k = (d_{k1}, d_{k2}, \ldots, d_{ks_k})$.

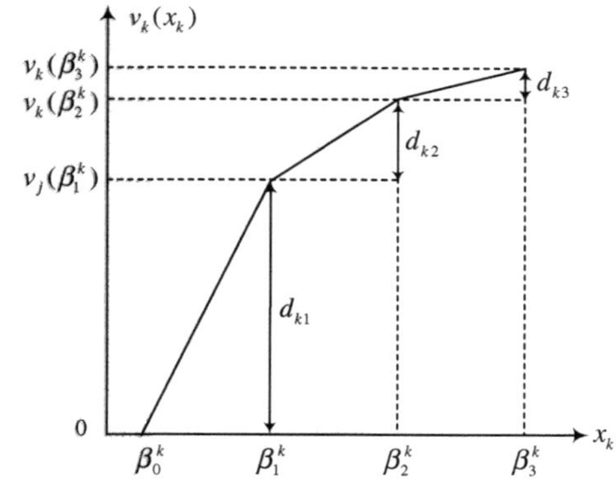

Fig. 4.2 Piecewise linear modeling of a marginal value function

The parameters of model (4.7) can be estimated in the context of the MCDA disaggregation paradigm [128] with non-parametric linear programming formulations, using data for obligors classified into predefined risk classes. Such data can be collected from historical data bases of financial institutions. Usually the data consist of defaulted and non-defaulted obligors, but multi-grading schemes are also possible, such as the credit ratings issued by credit rating agencies [69].

In a general setting, let us assume that reference (training) data for $M_1, M_2, \ldots, M_N$ obligors are available from N risk classes $C_1, \ldots, C_N$, defined such that C_1 is the low risk category and C_N the higher risk one. The decisions based on a credit scoring model $V(\mathbf{x})$ are made on the basis of the following classification rule:

$$\text{Obligor } i \text{ belongs in risk category } \ell \iff t_\ell < V(\mathbf{x}_i) < t_{\ell-1} \tag{4.8}$$

where $1 > t_1 > t_2 > \cdots > t_{N-1} > 0$ are score thresholds that distinguish the risk classes. The scoring model and thresholds that best fit the above rule, according to the available training data for M obligors can be estimated through the solution of the following linear programming problem [71]:

$$\begin{aligned}
&\min \quad \sum_{\ell=1}^{N} \frac{1}{M_\ell} \sum_{\mathbf{x}_i \in C_\ell} (\varepsilon_i^+ + \varepsilon_i^-) + \lambda \sum_{k=1}^{K} \mathbf{1}^\top \mathbf{d}_k \\
&\text{s.t.} \quad V(\mathbf{x}_i) = \sum_{k=1}^{K} \mathbf{p}_{ik}^\top \mathbf{d}_k \qquad i = 1, 2, \ldots, M \\
&V(\mathbf{x}_i) - t_n + \varepsilon_i^+ \geq 1, \qquad \forall \mathbf{x}_i \in C_\ell,\ \ell = 1, \ldots, N-1 \\
&V(\mathbf{x}_i) - t_{n-1} - \varepsilon_i^- \leq -1, \quad \forall \mathbf{x}_i \in C_\ell,\ \ell = 2, \ldots, N \\
&t_{\ell-1} - t_\ell \geq 0, \qquad \ell = 2, \ldots, N-1 \\
&\mathbf{d}_k, t_\ell, \varepsilon_i^+, \varepsilon_i^- \geq 0 \qquad \forall i, k, \ell
\end{aligned} \tag{4.9}$$

where **1** is a vector of ones. The first set of constraints defines the credit scores for the training cases according to the additive model (4.7). The second set of constraints defines the violations (ε^+) of the lower bound of each risk class (this applies only to obligors belonging to classes $C_1, \ldots, C_{N-1}$), whereas the third set of constraints defines the violations (ε^-) of the upper bound of each risk category (this applies only to the obligors belonging to classes $C_2, \ldots, C_N$). The last constraint is used to ensure that the thresholds are monotonically non-increasing.

The objective function combines two terms. The first involves the minimization of model's fitting error. This is defined as the weighted sum of the errors for cases belonging into different classes, where the weights are defined in terms of the number of sample observations in each class. In this way, it is possible to handle reference sets with considerable imbalanced class sizes, which are very common in credit scoring (e.g., the number of obligors is default is much lower than the non-defaulted obligors). The second term in the objective function is a regularization term in accordance with Tikhonov's regularization principle [242]. The parameter $\lambda > 0$ defines the trade-off between the minimization of the fitting error and the complexity of the model, which can be set by trial-and-error or with statistical resampling techniques such as such as cross-validation [233] and the bootstrap [79].

Denoting by $\mathbf{d}_k^*$ $(k = 1, \ldots, K)$ the optimal parameters of the model resulting from the solution of the above linear program, the constructed additive value function is scaled between zero and $\theta = \sum_{k=1}^{K} \mathbf{1}^\top \mathbf{d}_k$. Rescaling the model in [0, 1] can be easily done simply by dividing the optimal solution by θ.

The use of linear programming for model fitting enables the handing of big data sets. This is particularly important for credit scoring, as the available data become larger, particularly after the introduction of the Basel II regulatory framework. Furthermore, a linear programming model enables the risk analyst to incorporate special domain knowledge, which can be very useful for calibrating the model with expert judgment, in order to capture aspects of the problem not adequately covered by the data. Finally, post-optimality techniques can be employed to analyze the robustness of the results and the obtained model [75].

This modeling framework is also applicable with other types of decision models for credit scoring and alternative optimization techniques for model fitting. For instance, Doumpos [63] presented an evolutionary algorithm for constructing a non-monotone value function, whereas Doumpos and Zopounidis [74] used a similar algorithm for an outranking model. Bugera et al. [37] introduced goal programming models for developing a credit scoring model in the form of a quadratic value function, whereas Doumpos et al. [65] used the MHDIS method [266], which is based on multiple additive value models. Other optimization formulations (linear and non-linear) for fitting multicriteria credit scoring models have been also been proposed in several studies [99, 113, 197, 263].

Table 4.1 Number of sample observations in each year and category

Years	Non-defaulted	Defaulted	Total
2007	2,748	252	2,800
2008	2,846	253	2,899
2009	2,731	299	2,830
2010	2,143	244	2,187
Total	10,468	248	10,716

4.5 An Application

4.5.1 Data

To illustrate the usefulness and performance of MCDA methods in credit scoring, a large sample of Greek firms from the commercial sector (wholesale and retail trade) is used. The sample is taken from the database of ICAP S.A., which is a leading business information and consulting firm in Greece. The data span the period 2007–2010. In each year throughout that period, the firms in the database were classified either in the default or in the non-default group. The default group consists of firms that declared bankruptcy as well as firms with other default events such as protested bills, uncovered cheques, payment orders. Table 4.1 presents the number of observations from the two groups for each year in the sample.

The firms in the sample are described over seven financial ratios (Table 4.2), which cover three main aspects of corporate performance in accordance with the framework of Courtis [53]:

- *Profitability*: Profitability ratios assess the ability of a firm to generate earnings. The profitability ratios considered in this study include the gross profit margin (gross profit/sales) and return on assets (earnings before interest and taxes/total assets). The gross profit margin ratio is used to assess the sales profitability of the firms, after controlling for the cost of sales, whereas the return on assets ratio provides an overall evaluation of the operating profitability of the firms, taking into consideration all types of operating expenses.
- *Solvency and liquidity*: Solvency assesses the dependency of the firms on debt financing and their overall level of leverage. Liquidity, on the other hand, determines a company's ability to pay off its short-term debt obligations. In this study, total liabilities/total assets is used to assess the firms' solvency, whereas the liquidity of the firm is considered through the current ratio (current assets/current liabilities).
- *Managerial performance*: Managerial performance ratios focus on the efficiency of a firm's policies towards its creditors and clients as well as its financial efficiency. The former dimension is taken into consideration through two ratios, namely the receivables turnover ratio (accounts receivables × 365/sales) and the sales to current liabilities ratio. On the other hand, financial efficiency is analyzed through the interest expenses ratio (interest expenses/sales).

Table 4.2 Financial ratios for credit risk assessment

Financial ratios	Abbreviation	Relationship to credit risk
Profitability		
Gross profit/sales	GP/S	–
Earnings before interest and taxes/total assets	EBIT/TA	–
Solvency and liquidity		
Total liabilities/total assets	TL/TA	+
Current assets/current liabilities	CA/CL	–
Managerial performance		
Accounts receivables × 365/sales	AR/S	+
Sales/current liabilities	S/CL	–
Interest expenses/sales	IE/S	+

The risk of default increases with ratios that are positively related to credit risk

Table 4.3 Averages and standard deviations (in parentheses) of the financial ratios for each group of firms

	Non-default	Default
GP/S	0.30 (0.20)	0.23 (0.20)
EBIT/TA	0.04 (0.12)	−0.04 (0.14)
TL/TA	0.72 (0.27)	0.88 (0.25)
CA/CL	1.67 (1.55)	1.22 (1.07)
AR/S	237.31 (247.79)	342.55 (371.90)
S/CL	2.57 (2.96)	1.51 (2.41)
IE/S	0.03 (0.04)	0.07 (0.08)

The selection of the financial ratios was based on the combination of three main factors: (a) the research literature and the current best practices in the area of credit scoring by international organizations, (b) the judgment of credit scoring analysts with significant expertise on the characteristics of Greek firms, and (c) the discriminating power of the ratios.

Table 4.3 provides some basic statistics for the selected financial ratios for each group of firms. As expected, firms in default have lower profitability, higher debt burden and lower liquidity, higher interest expenses, and are less efficient in terms of the credit they provide to their clients (accounts receivable turnover) and the management of their short-term liabilities (S/CL ratio). The differences between the two groups are all found to be significant at the 1 % level through the Mann-Whitney non-parametric test.

4.5.2 Results

In order to be able to assess the predictive performance of a credit scoring model, a holdout sample is required, ideally involving data for different obligors and time period compared to the data set used to construct the scoring model [228]. In that

Table 4.4 Contribution of the financial rations in the model

	UTADIS	LR
GP/S	0.009	−0.684(0.068)
EBIT/TA	0.184	−3.645(0.195)
TL/TA	0.180	1.791 (0.207)
CA/CL	0.199	−0.332(0.158)
AR/S	0.104	0.001 (0.152)
S/CL	0.131	0.047 (0.069)
IE/S	0.194	8.113 (0.151)
Constant		−1.497

regard, the sample described in the previous section is split in two parts. The first covers the period 2007–2008 and it is used as the training sample, whereas the data for the period 2009–2010 are employed to test the performance of the credit scoring models (i.e., holdout sample).

Following this procedure, Table 4.4 reports the trade-offs of the financial ratios in the multicriteria additive model, as estimated through the solution of the linear program (4.9) using the 2007–2008 data. For comparative purposes the coefficients of the ratios in a logistic regression (LR) model are also reported, together with their relative contribution in the model (in parentheses). LR is the most popular statistical approach for constructing credit scoring models and it is widely used in this field by both researchers and practitioners. In the context of LR, the relative importance of the ratios can be assessed through the following index:

$$w_k = \frac{|\alpha_k|\sigma_k}{\sum_{k=1}^{K} |\alpha_k|\sigma_k}$$

where α_k is the regression coefficient of ratio k and σ_k is the standard deviation of the ratio. Measured in this way, w_k represents the relative influence of ratio k on the LR result in terms of the absolute impact of a standard deviation change in the ratio as a proportion of the total absolute change in the dependent variable, given a standard deviation change in all ratios [2].

In the multicriteria model, liquidity (CA/CL), interest expenses (IE/S), return on assets (EBIT/TA), and solvency (TL/TA) have the highest trade-offs and consequently they are important factors in the credit scoring process. The same variables also contribute significantly in the LR model (all coefficients are significant at the 1 % level). However, it is worth noting that the coefficient of the sales/short-term liabilities ratio has an incorrect sign in the LR model, as its positive regression coefficient indicates that the probability of default increases with this ratio, which does not comply with the economic interpretation of this ratio.

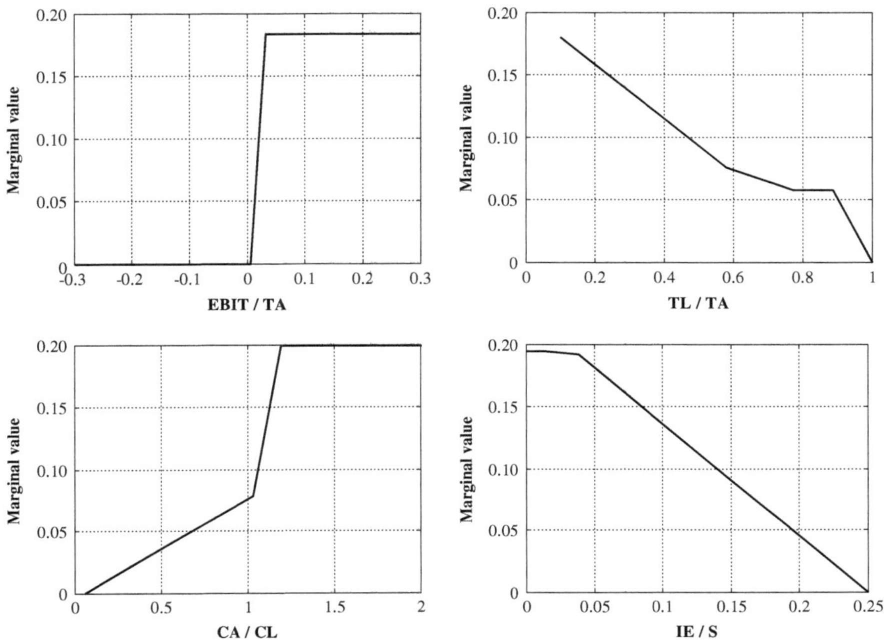

Fig. 4.3 Marginal value functions

Figure 4.3 illustrates the marginal value functions of the four most significant ratios in the multicriteria model. These functions provide further insights into how the overall credit score of the firms is affected by their performance on these ratios. For instance, EBIT/TA has a type of a step function, with the marginal value (partial credit score) increasing for positive values of the ratio. Thus, the likelihood of default is significantly lower for firms with positive return on assets. A similar behavior is also observed for liquidity; the credit score increases (improves) linearly when CA/CL is below one, but improves significantly for higher values. On the other hand, the marginal value functions for the solvency and the interest expenses ratios have a nearly linear form. As far as the TL/TA ratio is concerned, the partial credit score decreases almost linearly for firms with TL/TA < 0.9, but it is significantly lower for firms facing a higher debt burden. On the other hand, the marginal value for the IE/S ratio remains at high levels for firms with IE/S lower than 5 % and decreases linearly for firms with higher interest expenses. This kind information derived from the marginal value functions of the credit assessment criteria can be of great help for credit analysts, as it enables them to have a better understanding of the credit scoring model.

Except for analyzing the structure of a credit scoring model and the role of the credit assessment criteria, the relationship between the probability of default and the credit scores of the model as well as the model's predictive performance, are also

Fig. 4.4 Distribution of sample observations in the credit rating classes

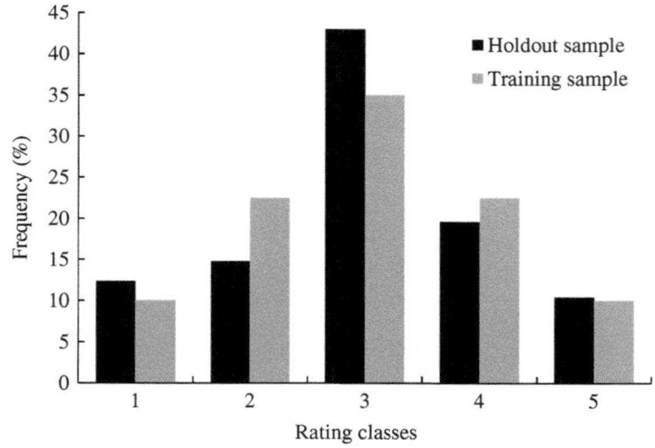

critical issues for the implementation of the model in practice. In this application, these issues are analyzed by applying the constructed multicriteria model to the 2009–2010 holdout data. In order to examine the relationship between the probability of default and the credit scores of the model, the latter are mapped to a five-point credit rating scale. The rating scale is defined on the basis of the global values (credit scores) of the observations in the training sample, as follows:

- class 1: very low risk firms with $V(\mathbf{x}_i) \geq 0.899$,
- class 2: low risk firms with $0.82 \leq V(\mathbf{x}_i) < 0.899$,
- class 3: medium risk firms with $0.617 \leq V(\mathbf{x}_i) < 0.82$,
- class 4: high risk firms with $0.513 \leq V(\mathbf{x}_i) < 0.617$,
- class 5: very high risk firms with $V(\mathbf{x}_i) < 0.513$.

The thresholds are set such that the firms are approximately normally distributed in the five rating classes, according to the available data for the calibration of the model (i.e., the training data). In that regard, the top 10 % of the training cases are assigned to class 1 (i.e., the threshold 0.899 is the 90 % percentile of the global values of the training observations). The next 22.5 % of the training cases are assigned to class 2 (i.e., the threshold 0.82 is the 67.5 % percentile of the global values of the training observations). Following the same approach, the threshold 0.617 that distinguishes medium risk firms from high risk ones, corresponds to the 32.5 % percentile of the scores in the training sample (i.e., the medium risk category consists of 35 % of the cases), whereas the threshold 0.513 that distinguishes high risk firms from very high risk ones corresponds to the 10 % percentile of the scores in the training sample (the high risk group includes 22.5 % of the training observations and the very high risk class includes the bottom 10 %). The score thresholds specified in this way are then used to rate the firms in the holdout sample. As shown in Fig. 4.4, the distribution of the sample observations in the five rating classes exhibits good stability when comparing the results for the training and holdout samples.

On the basis of this five-point rating, the probability of default in each rating class can be estimated as the number of cases in default to the total number of cases in each

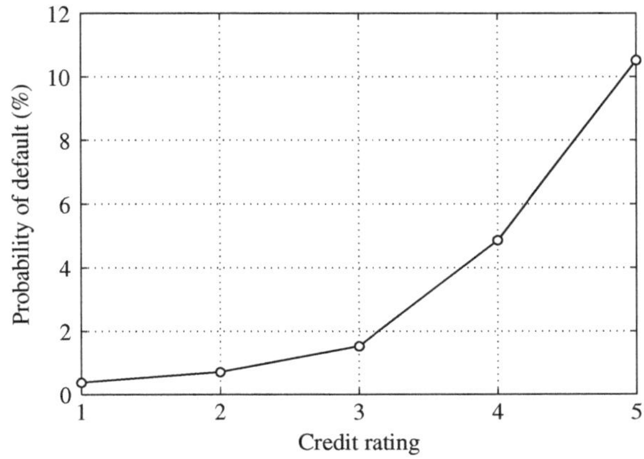

Fig. 4.5 Empirical probability of default for each rating class (holdout sample)

category. Figure 4.5 illustrates the results for the holdout sample. It is evident that the risk of default increases exponentially when moving from low risk grades to high risk ones (by a factor of about two). This is an appealing feature for a credit scoring model, as it indicates that the model provides a clear differentiation of the obligors in terms of their risk level, and its results are in accordance with the empirical default frequency in the data.

To further analyze the predictive ability of the multicriteria model, different performance measures are employed:

- *Accuracy rates*: On the basis of the credit scores estimated through a model and a cut-off point, obligors are classified in the pre-defined risk categories (default and non-default). Then, different accuracy measures can be defined. In this application we use two main accuracy criteria:
  - *Overall classification accuracy* (*OCA*): the ratio between the model's correct classifications to the total number of obligors evaluated. Similar calculations can be made (separately) for each risk category. Thus, the accuracy rate α_{ND} for the non-default group is defined as the percentage of non-defaulted obligors classified correctly by the model. The accuracy rate α_D for the default group is defined in the same way (i.e., the percentage of defaulted obligors classified by the model in the default category).
  - *Average classification accuracy* (*ACA*): the average of α_{ND} and α_D. This averaging can be justified for the most common setting where a credit scoring model is constructed with data for defaulted and non-defaulted obligors, taking into account the expected misclassification cost. In particular, denoting by p_D the a-priori probability of default, the expected misclassification cost of a credit scoring model is:

$$E(C) = p_D C_D \alpha_D + (1 - p_D) C_{ND} D a_{ND},$$

where C_D is the cost of misclassifying an obligor in default and C_{ND} is the cost for a non-defaulted obligor. The former is associated with losses due to default, whereas the latter is related to the opportunity cost derived by rejecting credit to a creditworthy client. Obviously C_D is much higher than C_{ND}, but on the other hand, the a-priory probability of default (p_D) is generally low (e.g., typically around 5 %). Thus, it is reasonable to assume that $p_D C_D \approx (1-p_D)C_{ND} = P$, in which case:

$$E(C) \approx 2P\frac{\alpha_{ND}+\alpha_D}{2} = 2P \times ACA$$

Thus, ACA is (in general) a reasonable proxy for the expected cost that arises from using a credit scoring model.

- *Area under the receiver operating characteristic curve*: (*AUROC*) The AUROC provides an overall evaluation of the generalizing performance of a classification model without imposing any assumptions on the misclassification costs or the prior probabilities [87] and it is commonly used to assess the discriminating power of credit rating models [30, 82, 222]. Formally, the AUROC represents the probability that a non-defaulted obligor will receive a higher credit score compared to one in default. Thus, it can be calculated as follows:

$$\text{AUROC} = \frac{1}{M_D M_{ND}} \sum_{i \in ND} \sum_{j \in D} I(\mathbf{x}_i, \mathbf{x}_j)$$

where M_D, M_{ND} denote the number of observations in default and non-default respectively and $I(\mathbf{x}_i, \mathbf{x}_j)$ is defined for a credit scoring model $V(\mathbf{x})$ as follows:

$$I(\mathbf{x}_i, \mathbf{x}_j) = \begin{cases} 1 & \text{if } V(\mathbf{x}_i) > V(\mathbf{x}_j) \\ 0.5 & \text{if } V(\mathbf{x}_i) = V(\mathbf{x}_j) \\ 0 & \text{if } V(\mathbf{x}_i) < V(\mathbf{x}_j) \end{cases}$$

- *Kolmogorov-Smirnov distance*: (*KS*) The Kolmogorov-Smirnov distance is the maximum absolute difference between the cumulative distribution functions of the credit scores of the obligors belonging into different groups. The highest is this difference the more powerful is a credit scoring model in discriminating the risk classes.

Table 4.5 presents detailed results on the predictive ability of the multicriteria credit scoring model according to the above performance measures. For comparison purposes, the results of LR are reported as well as those of a support vector machine (SVM) model developed with a radial basis function kernel (SVM-RBF) using the LIBSVM library in MATLAB R2013 [42]. SVMs have become an increasingly popular statistical learning methodology for developing classification and regression models [249] with many successful applications in financial decision-making

Table 4.5 Comparative results for the predictive performance of the models

Measures	Methods	2009	2010	2009–2010
α_{ND}	UTADIS	0.720	0.723	0.721
	LR	0.676	0.690	0.682
	SVM-RBF	**0.735**	**0.745**	**0.739**
α_D	UTADIS	**0.707**	**0.818**	**0.741**
	LR	0.697	0.795	0.727
	SVM-RBF	0.687	0.750	0.706
ACA	UTADIS	**0.713**	**0.771**	**0.731**
	LR	0.687	0.743	0.705
	SVM-RBF	0.711	0.748	0.723
OCA	UTADIS	0.719	0.725	0.722
	LR	0.677	0.692	0.683
	SVM-RBF	**0.733**	**0.745**	**0.738**
KS	UTADIS	**0.442**	**0.567**	**0.479**
	LR	0.393	0.515	0.429
	SVM-RBF	0.435	0.512	0.453
AUROC	UTADIS	**0.769**	**0.826**	**0.786**
	LR	0.756	0.815	0.775
	SVM-RBF	0.767	0.825	0.785

The best result for each performance measure is marked in bold

problems, including credit scoring [23, 123, 171, 234]. The use of the RBF kernel enables the development of nonlinear classification models, as opposed to the linear modeling setting of LR and the additive nature of the MCDA approach used in this analysis.

The results of Table 4.5 indicate that the multicriteria credit scoring model consistently outperforms LR on all performance measures and time periods, while being quite competitive to the SVM-RBF nonlinear model. In particular, the UTADIS model outperforms LR and SVM-RBF in identifying firms in default. Throughout the two-years period 2009–2010, the accuracy of the multicriteria model for the firms in default is 74.1 % versus 72.7 and 70.6 % for LR and SVM-RBF. On the other hand, the SVM-RBF model performs better for the non-default group, whereas LR performs poorly compared to the other methods. Overall, the UTADIS model achieves the best balance between the accuracy rates for the two risk groups. As a result, it outperforms the other models in terms of ACA (73.1 % overall versus 72.3 % for the SVM-RBF models, and 70.5 % for the LR model). The good performance of the SVM-RBF model for the non-default group (which is the largest one; cf. Table 4.1) leads to its high OCA. The UTADIS model follows second in terms of its OCA. Finally, as far as the two performance measures that do not involve accuracy rates are concerned, the UTADIS model performs consistently better than LR and SVM-RBF. The differences are higher for the KS distance, whereas in terms of AUROC the multicriteria model and SVM-RBF perform almost equally well.

Chapter 5
Portfolio Management

Abstract Portfolio management is often viewed as a bi-criteria risk-return optimization problem in accordance with the well-known mean-variance framework. In this chapter, portfolio management is considered in a broader context, covering asset selection and portfolio optimization taking into consideration different risk-return measures and other goals and objectives, which are commonly used by investors and portfolio managers. Extensions of this framework to mutual funds and index tracking are also discussed, together with implementations in decision support systems.

Keywords Portfolio management · Multiobjective optimization · Asset selection · Index tracking · Mutual funds

5.1 The Portfolio Management Process

Portfolio management has been one of the major fields in finance since the introduction of the mean-variance model by Markowitz [167]. Portfolio management is involved with the construction of portfolios of securities (stocks, bonds, treasury bills, mutual funds, etc.) that maximize the investor's utility. Generally, this can be considered as a dynamic and ongoing process implemented in three main stages as illustrated in Fig. 5.1.[1]

The first stage of the process involves the problem structuring phase, which ultimately leads to the selection of suitable investment options for the investor. The analysis at this stage takes into consideration a number of factors and decision criteria, depending on the investment preferences of the investor and the existing options under consideration. Typically, these involve fundamental factors, such as the conditions and trends in the macroeconomic environment, data about specific business sectors, the characteristics of the available investment options, as well as technical

[1] A similar modeling process is also described by Maginn et al. [160], who refer to the three stages as planning, execution, and feedback.

M. Doumpos and C. Zopounidis, *Multicriteria Analysis in Finance*,
SpringerBriefs in Operations Research, DOI: 10.1007/978-3-319-05864-1_5,

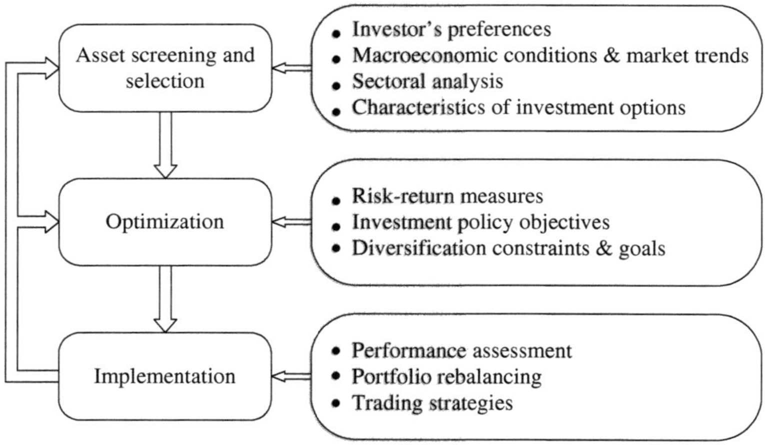

Fig. 5.1 The portfolio management process

analyses to examine the trends and sentiment in the financial markets. The outcome of the first stage is a limited set of selected assets (from one or multiple classes such as stocks, bonds, mutual funds, etc.), which are candidates to be included in an investment portfolio.

The second stage focuses on the portfolio optimization process, which involves the allocation of the available capital to the assets selected at the first stage. The allocation is implemented in accordance with the investor's preferences with respect to the desired levels of return and risk as well as other goals/objectives and constraints that determine his/her investment policy, such as:

- regulatory issues (e.g., whether short-selling is allowed or not),
- transaction costs,
- liquidity considerations,
- diversification constraints and objectives,
- the investment's time horizon (e.g., single-period versus multi-period investments).

Finally, the last stage is involved with implementation issues related to the development of proper trading strategies, the monitoring and assessment of the portfolio's performance, and the revision of the portfolio's composition on the basis of the markets' dynamics and the changes in the investor's policy and objectives. Depending on the actions decided, the process may iterate from the previous stages.

In the following sections we discuss in more detail the phases of asset screening and portfolio optimization, focusing on their multicriteria aspects and illustrate MCDA techniques that can be used in these contexts. Xidonas et al. [257] provide a comprehensive analysis of the field, including a detailed discussion of the portfolio management process, an extensive overview of different methods, numerical illustrations, as well as a coverage of practical issues.

5.2 Asset Screening and Selection

Before proceeding to asset allocation decisions, traders and investors evaluate and select asset classes and specific assets within each class. In a passive management setting this is done periodically, whereas in active strategies such decisions are taken much more frequently. Asset screening and selection improves the potentials of risk diversification (especially when selecting assets from different classes) and reduces the complexity of the portfolio management process, by focusing only on assets with certain characteristics which are deemed as important by the investor.

In the context of financial theory, asset selection is based on single and multi-factor models, such as the capital asset pricing model [216], arbitrage pricing theory [205], and newer extensions and variants [40, 85, 86] (see Sect. 5.4.1). Together, with such models, portfolio managers consider a number of fundamental and technical analysis indicators. The former provide information on the medium and long-term prospects of the assets under consideration and the market, whereas technical analysis is employed to identify short-term trends in asset prices. For instance, in the case of equity portfolios, common fundamental indicators include stock market ratios (market to book value, price/earnings, earnings per share, dividends/earnings, etc.), financial ratios (profitability, solvency, liquidity, leverage, managerial performance), as well as indicators related to trends in the external economic environment. On the other hand, technical analysis is commonly based on moving averages, price and volume-based indicators, oscillators, etc.

MCDA approaches used to support the asset selection process have been based on MAVT and outranking models. Saaty et al. [211] first employed a MCDA approach for this purpose in the context of stock evaluation. In particular, the authors presented a hierarchical modeling approach based on the analytic hierarchy process. The proposed hierarchy was based on extrinsic and intrinsic factors, as well as the investor's objectives, resulting to a ranking of a set of stocks according to their performance on these three dimensions.

A similar hierarchical evaluation scheme was also employed by Bana e Costa and Soares [52], who followed a two-stage multicriteria process. In the first stage, a multicriteria model was built to assess the return dimension through a combination of multiple intrinsic and extrinsic factors. Similarly to the aforementioned study of Saaty, intrinsic factors were used to assess the performance of firms, on the basis of indicators such as dividend yield, earnings per share, price/earnings ratio, etc. The external factors, on the other hand, relate to the general economic conditions (GDP growth, interest rates, the effect of mergers/acquisitions, etc.) and to specific information about the business sector of the firms. The evaluation model was expressed in the form of an additive value function, constructed interactively with the investor. In a second stage, the multicriteria model was used to construct a portfolio of stocks through an optimization model.

Samaras et al. [213] also considered stock selection through multicriteria models expressed in the form of an additive value function. In particular, the authors presented an interactive decision support system that enables the user to build stock selection models using fundamental financial and stock market data, as well as qualitative information. The system also includes customized screening rules specifically calibrated for firms belonging into different business sectors (e.g., banks, insurance companies, industry, services, etc.).

Following a different approach, Hurson and Zopounidis [124] proposed a combination of a preference disaggregation technique with an outranking classification approach. The former was used to construct an additive value model based on reference data for a set of stocks assessed in terms of their attractiveness by an expert stock portfolio analyst. The model combined performance criteria based on market data, such as return, marketability, systematic risk, the price/earnings ratio, dividends, as well as two financial ratios related to the liquidity and profitability of the firms. The resulting additive model provided a ranking of the stocks from the most to the least promising ones. On the other hand, the outranking approach based on the ELECTRE TRI method [209] was used to enhance the results of the value model, by providing a classification of the stock into three homogeneous performance categories.

Such a combination of classification and ranking models is well-suited for asset screening and selection. The classification of assets into performance categories provides investors and portfolio managers with easy to use information by distinguishing promising investment choices from poor ones. On the other hand, the ranking of assets in each performance category enables the analysis of the relative performance of one investment option as opposed to others, and facilitates the asset screening process by providing an additional filter that can be used to further reduce the set of options to the top investments from each performance group.

A classification approach was also used by Zopounidis et al. [269] who employed two preference disaggregation techniques to build multicriteria models for the classification of stocks into three performance categories defined by an expert portfolio manager. The models combined 15 criteria, including valuation ratios, marketability, returns, dividend yield, and corporate performance financial ratios. The results from the application to a sample of Greek stocks, showed that indicators related to marketability were the most important, followed by the financial performance of the firms. The models classified the stocks accurately in the predefined categories, outperforming a discriminant analysis model.

Xidonas et al. [258], on the other hand, proposed an interactive classification approach based on the ELECTRE TRI outranking method. In contrast to previous studies, which have used a common set of selection criteria applicable to all business sectors, the authors suggested the consideration of different selection criteria for industry/commerce firms, financial services, banking institutions, and insurance companies. A validation of the modeling approach in terms of the risk-return performance of the selected stocks confirmed that the validity of multicriteria evaluation results.

5.3 Portfolio Optimization

5.3.1 The Mean-Variance Framework

The allocation of an available capital to a set of selected assets is a well-known optimization problem, which can be considered in different contexts (e.g., static/dynamic, deterministic/stochastic). The foundations of portfolio optimization have been set by Markowitz [167] through the introduction of the mean-variance (MV) model.

In the MV framework context, the objective of the investor is to identify *efficient portfolios* that minimize risk for a given level of expected return. Formally, the MV model can be formulated as a quadratic optimization problem of the following form:

$$\begin{aligned} \min \quad & \mathbf{x}^\top \mathbf{V}\mathbf{x} \\ \text{s.t.}: \quad & \bar{\mathbf{r}}^\top \mathbf{x} \geq R \\ & \mathbf{1}^\top \mathbf{x} = 1 \\ & \mathbf{a} \leq \mathbf{x} \leq \mathbf{b} \end{aligned} \tag{5.1}$$

where:

- $\mathbf{V}$ is a $M \times M$ matrix with the covariances of the assets returns,
- $\bar{\mathbf{r}} = (\bar{r}_1, \ldots, \bar{r}_M)$ is the vector with the expected returns of the assets,
- R is the desired level of expected return,
- $\mathbf{x} = (x_1, \ldots, x_M)$ is the vector with the weights of M assets in the portfolio,
- $\mathbf{a}$ and $\mathbf{b}$ are vectors with the lower and upper bounds for the proportion of capital invested in each asset,
- $\mathbf{1}$ is a vector of ones.

Solving the quadratic program (5.1) for different values of R leads to different efficient portfolios, among which the investor/portfolio manager may chose the one that best meets his/her investment policy.

The MV approach is founded on the principles of expected utility theory. In particular, it is assumed that investors are risk-averse and their choices among different portfolios are driven by their expected returns and variances. These conditions imply that the investors' preferences are described by a quadratic utility function of return r, of the form $U(r) = a + br + cr^2$. As noted by Markowitz [165], the assumption of a quadratic utility function, or making assumptions about the normality of returns, are sufficient conditions for applying the MV framework, but they are not necessary. Markowitz argues that the MV frontier approximately maximizes expected utility for a wide class of concave (risk-averse) utility functions and overviews empirical results indicating that this holds for different distributions of returns. In such a setting, there is no need to explicitly define utility function of an investor, as a careful selection of a portfolio on the MV frontier will be approximately optimal.

5.3.2 Alternative Portfolio Selection Criteria

Despite the fundamental contributions of the MV framework in finance and decision theory, nowadays portfolio optimization is a much more involved process that requires a richer description of a portfolio's risk-return characteristics. Indeed, the frequent appearance of financial crises has highlighted the significance of tail risk, which arises from extreme losses in highly adverse conditions. This is further justified by the increasing interest in special investment types, such as hedge funds, which are intensionally designed to have non-normal distributions and significant downside risk [215]. As a consequence, the consideration of only the two first moments (mean, variance) of the returns distribution is often insufficient. Furthermore, departures from normality with regard to the distribution of returns make questionable the use of variance as a measure of dispersion.

In his 1959 book, Markowitz [167] actually discussed extensions of the MV approach, introducing (briefly) other risk measures such as semivariance, mean absolute deviation, and expected loss. Over the years, such measures have been operationalized through the introduction of new model formulations. Three popular models include:

- The semivariance (SV) model [166]: In contrast to considering both upward and downward deviations from mean return, the semivariance model only considers downward deviations. Formally, given a series of returns $r_{1i}, \ldots, r_{Ti}$ for the returns of an asset i over T time periods, SV is defined as:

$$SV_i = \frac{1}{T} \sum_{t=1}^{T} [\min\{r_{ti} - \bar{r}_i,\ 0\}]^2$$

  Denoting by $\mathbf{D} = [r_{ti} - \bar{r}_i]$ $(i = 1, \ldots, M,\ t = 1, \ldots, T)$ a matrix with the deviations of the assets' returns from their means over T time periods, a portfolio that minimizes SV can be constructed from the solution of the following quadratic program:

$$\begin{aligned} \min \quad & \mathbf{y}^\top \mathbf{y} \\ \text{s.t.}: \quad & \mathbf{D}\mathbf{x} - \sqrt{T}\mathbf{y} + \sqrt{T}\mathbf{z} = \mathbf{0} \\ & \bar{\mathbf{r}}^\top \mathbf{x} \geq R \\ & \mathbf{1}^\top \mathbf{x} = 1 \\ & \mathbf{a} \leq \mathbf{x} \leq \mathbf{b} \\ & \mathbf{y}, \mathbf{z} \geq \mathbf{0} \end{aligned} \tag{5.2}$$

- Mean absolute deviation model [143]: Substituting variance with the mean absolute deviation (MAD) criterion overcomes the normality assumption of the MV model. For an asset i with returns $r_{i1}, \ldots, r_{iT}$ over T periods, MAD is defined as the average of the absolute deviations $|r_{ti} - \bar{r}_i|$, $t = 1, \ldots, T$, i.e.:

$$MAD_i = \frac{1}{T}\sum_{t=1}^{T} |r_{ti} - \bar{r}_i, |$$

The construction of a portfolio that minimizes MAD for a given level of experted return R can be done through the following linear program:

$$\begin{array}{ll} \min & \mathbf{1}^\top \mathbf{y} \\ \text{s.t.}: & -\mathbf{y} \le \mathbf{Dx} \le \mathbf{y} \\ & \bar{\mathbf{r}}^\top \mathbf{x} \ge R \\ & \mathbf{1}^\top \mathbf{x} = 1 \\ & \mathbf{a} \le \mathbf{x} \le \mathbf{b} \\ & \mathbf{y} \ge \mathbf{0} \end{array} \tag{5.3}$$

Similarly to the MV model, MAD can also be extended to focus on downside risk [176].

- Conditional value at risk [202]: Value at risk (VaR) has been widely used in many areas of financial risk management [135]. VaR_β is defined as the maximum expected loss over a specific time period with confidence level β. Despite its widespread use in practice, VaR has received much criticism as being an incoherent risk measure [9, 235]. Furthermore, from an optimization perspective, constructing portfolios that minimize VaR is computationally difficult [98]. Conditional VaR_β (CVaR) addresses these issues and has gained much interest. CVaR_β can be defined as the expected value of the losses exceeding VaR_β. Denoting by $\mathbf{R}$ a $T \times M$ matrix with asset returns over T time periods, the construction of portfolios that minimize this risk measure can be accomplished through the following linear program:

$$\begin{array}{ll} \min & \alpha + \frac{1}{(1-\beta)T}\mathbf{1}^\top \mathbf{y} \\ \text{s.t.}: & \alpha + \mathbf{y} + \mathbf{Rx} \ge \mathbf{0} \\ & \bar{\mathbf{r}}^\top \mathbf{x} \ge R \\ & \mathbf{1}^\top \mathbf{x} = 1 \\ & \mathbf{a} \le \mathbf{x} \le \mathbf{b} \\ & \mathbf{y} \ge \mathbf{0}, \alpha \in \mathbb{R} \end{array} \tag{5.4}$$

Kolm et al. [142] provide a comprehensive overview of the developments the area of portfolio optimization over the past six decades and describe several extensions of the basic MV framework, including new risk measures, optimization techniques, and practical issues, while Mansini et al. [162] focus on optimization models and risk measures based on linear programming. Another survey by Fabozzi et al. [83], highlights the variety of risk measures and approaches, which are currently used in practice, emphasizing the need to "to merge the different risk views into a coherent risk assessment". In this context, it is not surprising that the synergies among various approaches towards building a more comprehensive portfolio selection framework have already attracted much interest.

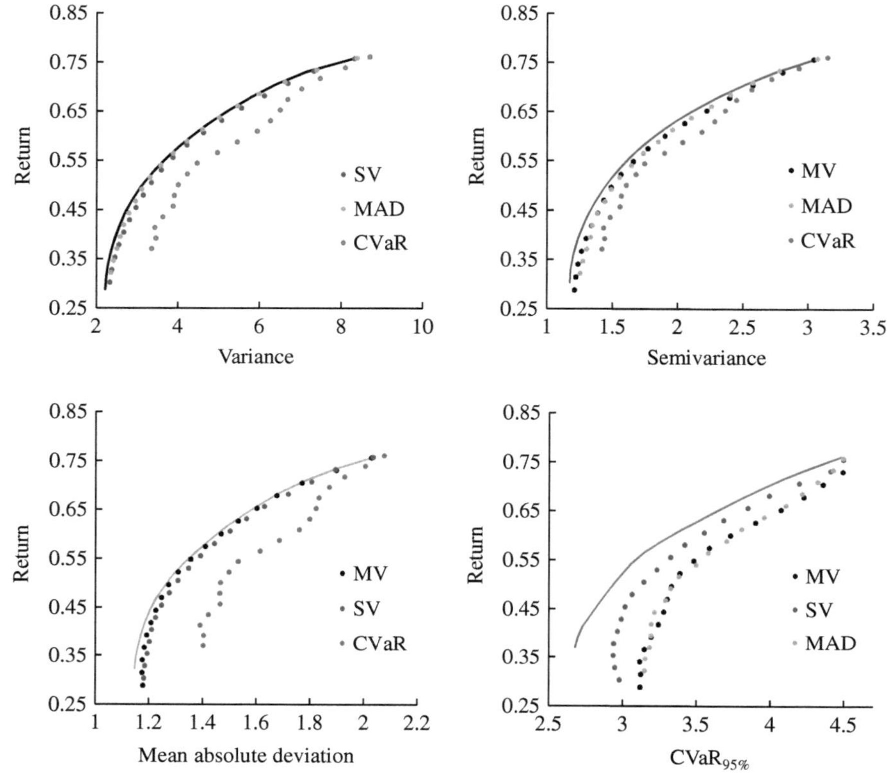

Fig. 5.2 Two-dimensional efficient frontiers constructed with different portfolio risk measures

As an example, Fig. 5.2 presents illustrative efficient frontiers constructed with the four optimization models described above, using weekly data for 30 companies in the Dow Jones Industrial Average, over the period 2011–2013. Each graph compares the frontier constructed with one risk measure to the results of the other models. For instance, in the return-variance graph, the black line corresponds to the MV frontier, whereas the dots illustrate the MV performance of portfolios optimized with other risk measures. It is evident that MV, MAD, and SV provide similar results, but the results obtained with CVaR (at the 95 % level) differ significantly from the other risk measures.

5.3.3 Multiobjective Portfolio Optimization

In the context described in the previous section, the consideration of the portfolio optimization process in a multiobjective context is particularly appealing, as it enables multiple risk measures and selection criteria to be incorporated in the analysis, in accordance with the investment preferences of a particular investor or portfolio manager.

The research on the use of multiobjective optimization and goal programming in portfolio optimization has grown significantly over the past decades, as computational and algorithmic advances now enable the consideration of different types of portfolio selection criteria. Most studies have focused on the combination of multiple measures that describe the risk-return properties of portfolios. Some examples include combinations involving skewness and kurtosis [35, 36, 58, 133, 140, 152, 261] value at risk and conditional value at risk [148, 161, 204, 247], mean absolute deviation [187], and systematic risk [203, 259].

For instance, portfolio optimization under return, variance, and CVaR objectives can be expressed in the form of the following multiobjective optimization problem[204]:

$$\begin{aligned} \max \quad & f_1(\mathbf{x}) = \bar{\mathbf{r}}^\top \mathbf{x} \\ \min \quad & f_2(\mathbf{x}) = \mathbf{x}^\top \mathbf{V}\mathbf{x} \\ \min \quad & f_3(\mathbf{x}) = \alpha + \frac{1}{(1-\beta)T}\mathbf{1}^\top \mathbf{y} \\ \text{s.t.} \quad & \alpha + \mathbf{y} + \mathbf{R}\mathbf{x} \geq \mathbf{0} \\ & \mathbf{1}^\top \mathbf{x} = 1 \\ & \mathbf{a} \leq \mathbf{x} \leq \mathbf{b} \\ & \mathbf{y} \geq \mathbf{0}, \alpha \in \mathbb{R} \end{aligned} \tag{5.5}$$

Given the convex nature of the problem, it can be easily solved by combining the three objectives into a weighted average performance measure $w_1 f_1(\mathbf{x}) - w_2 f_2(\mathbf{x}) - w_3 f_3(\mathbf{x})$. With this specification, different efficient portfolios can be constructed by varying the trade-off constants $w_1, w_2, w_3 \geq 0$, such that they sum up to one.

However, this approach will not work well in portfolio optimization under nonconvex objectives and/or constraints. For instance, often the number of assets included in the final portfolio is an important issue, as portfolios of too many assets are difficult to monitor and manage. Thus, small portfolios are usually preferred by investors and portfolio managers. Taking this issue into consideration, leads to cardinality constrained portfolios, which consist of a predefined maximum number of assets. Cardinality constraints in a portfolio optimization model, can be formulated by introducing binary decision variables $\mathbf{q} = (q_1, \ldots, q_M)$ to indicate whether an asset i is included in the portfolio ($q_i = 1$) or not ($q_i = 0$) and adding the following constraints:

$$\begin{aligned} \mathbf{a}\mathbf{q} \leq \mathbf{x} \leq \mathbf{b}\mathbf{q} \\ \mathbf{1}^\top \mathbf{q} \leq U \end{aligned} \tag{5.6}$$

where U is the predefined maximum number of assets in the portfolio. Figure 5.3 presents an illustrative return-CVaR frontier containing at most three stocks, constructed using data from the Greek stock exchange. In contrast to the illustrations

Fig. 5.3 A non-convex efficient frontier with cardinality constraints

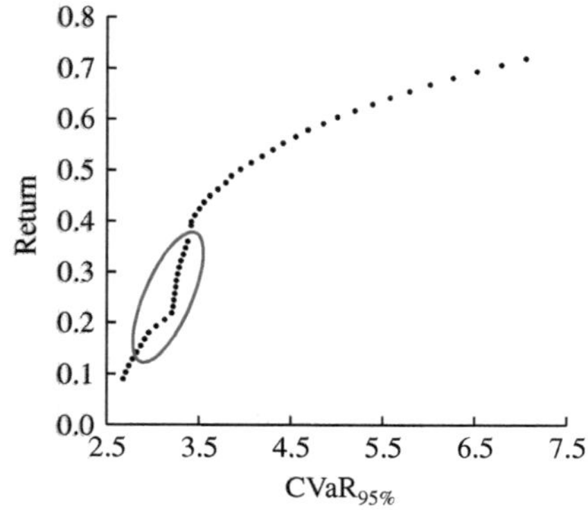

in Fig. 5.2, one can observe that the cardinality constrained frontier is not necessarily convex. In such cases, using a simple weighted aggregation of the portfolio optimization criteria could lead to loss of information, as it would be impossible to identify the portfolios indicated in the circled area.[2]

Addressing such issues can be easily done through alternative multiobjective and goal programming formulations. For instance, using the Chebyshev scalarization model described in Sect. 2.3.1, a mean-variance-CVaR$_\beta$ frontier can be constructed through the solution of the following optimization formulation:

$$\min \quad \gamma - \rho f_1(\mathbf{x}) + \rho f_2(\mathbf{x}) + \rho f_3(\mathbf{x}) \tag{5.7}$$

$$\text{s.t.} \quad f_1(\mathbf{x}) = \bar{\mathbf{r}}^\top \mathbf{x} \tag{5.8}$$

$$f_2(\mathbf{x}) = \mathbf{x}^\top \mathbf{V} \mathbf{x} \tag{5.9}$$

$$f_3(\mathbf{x}) = \alpha + \frac{1}{(1-\beta)T} \mathbf{1}^\top \mathbf{y} \tag{5.10}$$

$$\gamma \geq w_1[f_1^* - f_1(\mathbf{x})] \tag{5.11}$$

$$\gamma \geq w_2[f_2(\mathbf{x}) - f_2^*] \tag{5.12}$$

$$\gamma \geq w_3[f_3(\mathbf{x}) - f_3^*] \tag{5.13}$$

$$\alpha + \mathbf{y} + \mathbf{R}\mathbf{x} \geq \mathbf{0} \tag{5.14}$$

$$\mathbf{1}^\top \mathbf{x} = 1 \tag{5.15}$$

$$\mathbf{a} \leq \mathbf{x} \leq \mathbf{b} \tag{5.16}$$

$$\mathbf{y}, \gamma \geq \mathbf{0}, \alpha \in \mathbb{R} \tag{5.17}$$

The first three constraints define the three objectives (return, variance, CVaR$_\beta$), whereas constraints (5.11–5.13) define the maximum deviation from the ideal values of the objectives (maximum return—f_1^*, minimum variance—f_2^*, minimum CVaR$_\beta$—f_3^*). The remaining constraints define the set of feasible solutions in accor-

[2] In this example, there is no weighted average of the form w_1(Return) $-$ w_2(CVaR$_{95\,\%}$) with $w_1, w_2 \geq 0$, that would lead to the selection of any of the portfolios in the circled area.

dance with the formulations shown above. The full efficient frontier can be traced by varying the weighting constants w_1, w_2, w_3 associated with the three objectives.

An alternative approach to the construction of the full frontier of efficient portfolios is the ε-constrained formulation, in which one of the objectives is optimized, subject to constraints involving the others. For instance, the ε-constrained model for the mean-variance-CVaR$_\beta$ frontier can be formulated as follows:

$$\begin{aligned}
\min\ & \mathbf{x}^\top \mathbf{V}\mathbf{x} \\
\textbf{s.t.}\ & \bar{\mathbf{r}}^\top \mathbf{x} \geq R \\
& \alpha + \frac{1}{(1-\beta)T}\mathbf{1}^\top \mathbf{y} \leq C \\
& \alpha + \mathbf{y} + \mathbf{R}\mathbf{x} \geq \mathbf{0} \\
& \mathbf{1}^\top \mathbf{x} = 1 \\
& \mathbf{a} \leq \mathbf{x} \leq \mathbf{b} \\
& \mathbf{y} \geq \mathbf{0}, \alpha \in \mathbb{R}
\end{aligned}$$

where C is the maximum acceptable level for the portfolio CVaR. The full efficient frontier can be traced by parametrically varying C and R. An advantage of such a model compared to the Chebyshev scalarization formulation is that the construction of the frontier is based on parameters that an investor/portfolio manager can easily understand (e.g., the acceptable levels of CVaR and return). On the other hand, the Chebyshev uses weights for the objectives, which may not be straightforward to use. Furthermore, different weights can lead to the same or very similar results.

However, it should be noted that the use of different risk measures (separately or in combination) does not fully resolve the limitations and issues often encountered in risk-return portfolio optimization. For instance, such optimization models often exhibit instability with respect to their statistical inputs (i.e., risk-return estimates), the resulting portfolios may not be well diversified, and their composition may be non-intuitive (e.g., some assets having extreme weights) [142]. Most importantly, however, the concept of risk can not be adequately defined on the basis of past return distributions, as it depends on how an investor understands the risk in the investment period [177].

Therefore, for a portfolio optimization model to be practically useful, additional issues related to the results of the investment and the portfolio management process should be considered, in combination with sound investment judgment. Such issues include among others, transaction costs [261], liquidity considerations [155, 247], and dividends [259], regulatory constraints, as well as the investment horizon (e.g., dynamic portfolio optimization). The relevance of such additional portfolio selection criteria has been highlighted among others by Michaud [178] and more recently by Steuer et al. [231, 232], whereas Kolm et al. [142] overview the recent developments in portfolio optimization under such issues.

Furthermore, recently new trends have also emerged with regard to non-financial dimensions of portfolio management. A particularly interesting emerging field involves socially responsible investments (SRI), thus adding ethical, social, and environmental criteria in the portfolio construction process. According to the Forum of Sustainable and Responsible Investment, SRI investments in the USA stand at $3.74 trillion, covering about 10 % of the USA investment marketplace.[3] Hallerbach et al. [110] presented a framework for introducing SRI objectives into the portfolio selection process, whereas other studies have proposed multiobjective optimization and goal programming models combining traditional portfolio optimization measures with non-financial SRI criteria [15, 27, 28]. In a recent study Utz et al. [248] found out that socially responsible portfolios exhibit lower return volatility as socially responsible firms are generally less prone to earnings shocks. On the other hand, their results further show that portfolio managers use SRI criteria mainly at the asset selection stage rather than for asset allocation purposes.

Of course, the consideration of such multiple portfolio selection criteria, makes the portfolio optimization process much more involved. However, it provides a much more realistic framework that is close to the actual portfolio selection process as implemented by portfolio managers. At the same time, the effect of statistical errors is controlled by promoting the role of expert investment judgments.

On the methodological side, the added complexity due to the introduction of multiple portfolio objectives of diverse nature, can now be addressed with the computational and algorithmic advances. Several multiobjective approaches, formulations, and solution algorithms can be used for this purpose. As far as the problem formulations are concerned one can mention:

- multiobjective linear and nonlinear programming [161, 187, 204],
- goal programming [7, 10, 152],
- compromise programming [16, 35],
- stochastic programming [14, 25, 247], and
- fuzzy models [27, 251].

As far as the solution techniques are involves, these include traditional interactive methods [203] and parametric programming [116], as well as new algorithmic techniques that have recently attracted much interest in the context of portfolio optimization, such as multiobjective evolutionary algorithms [59, 148, 175] and metaheuristics [81]. These computational intelligence approaches have contributed significantly by allowing the handling of complex non-linear and non-convex portfolio selection criteria as well as the consideration of realistic diversification constraints (e.g., cardinality constraints), which as often hard to handle with analytic techniques in large-scale problems.

[3] http://www.ussif.org/sribasics

5.4 Mutual Funds Performance Appraisal

Mutual funds are probably the most widely used type of financial investment in modern financial markets. Their success is due to the unique advantages that they offer to investors, such as access to professional management. According to the Investment Company Institute, global investments in mutual funds have almost tripled during the last decade, reaching \$28.9 trillion at the end of the third quarter of 2013 compared to \$11.87 trillion in 2000. Most interestingly, the number of offered funds has steadily increased through time, exceeding 75,000 at the end of third quarter of 2013.

In light of the plethora of available funds, their evaluation and selection is a very challenging task. The financial theory provides an arsenal of measures for assessing fund performance, mainly based on the analysis of the funds' risk-return characteristics, as well as the market-timing, and selection abilities of fund managers. The following section provides a brief overview of the most commonly used measures.

5.4.1 Risk-Adjusted Performance Measures

The first performance measures to be proposed for assessing the performance of mutual funds, included risk-adjusted performance criteria, such as the Sharpe [217] and Treynor ratios [240]. Both these measures are ratios between a fund's excess return (r) over the risk-free rate (r_f) compared to risk, which is defined as the volatility (σ_E) of excess returns in the Sharpe ratio and systematic risk (β) in the ratio of Treynor.

$$\text{Sharpe ratio} = \frac{r - r_f}{\sigma_E} \qquad \text{Treynor ratio} = \frac{r - r_f}{\beta}$$

These performance criteria, however, do not distinguish between funds that systematically deliver superior risk adjusted returns due to their investment management efficiency, and those that just exploit market inefficiencies.

Thus, evaluation models have been developed to address the market-timing and stock selection abilities of fund managers. Jensen [129] introduced a regression model of a fund's excess returns against the excess returns of a benchmark market portfolio:

$$r - r_f = \alpha + \beta(r_M - r_f)$$

where r_M is the return of the market portfolio. The constant term α will be positive if the fund manager has any forecasting ability and zero in case of no forecasting ability. Similarly to the Treynor ratio, the Jensen's measure also assumes that the funds are well diversified, thus making systematic risk the only relevant risk dimension.

Other performance measures seek to asses both the market timing and security selection abilities of fund managers. For instance, Treynor and Mazuy [245] proposed the introduction of the additional quadratic term $Z = (r_M - r_f)^2$ to Jensen's regression, in order to test for market timing skills, whereas Henriksson and Merton [114] proposed instead the use of the term $Z = \max(0, r_M - r_f)$. With such additional terms, the original model of Jensen is expressed as

$$r - r_f = \alpha + \beta(r_M - r_f) + \gamma Z,$$

where α is a measure of a fund manager's stock selection ability and γ indicates the manager's market-timing ability.

However, traditional performance measures that rely on the mean-variance framework and the capital asset pricing model have received a lot of criticism. As a result, multi-factor models that incorporate additional risk factors in accordance with the framework of the arbitrage pricing theory [205]. The most representative examples of this line of research, are the models of French and French [85] as well as the model of Carhart [40]. French and French proposed the extension of the Jensen's model with two new factors that control for excess returns generated by tactical asset allocation strategies. These new factors involve: (a) the difference between the return on a portfolio of small capitalization stocks over a portfolio of big ones (*SMB*), and (b) the difference between the returns on a portfolio of stocks with high book-to-value ratio and a portfolio of low book-to-value stocks (*HML*):

$$r - r_f = \alpha + \beta(r_M - r_f) + s \cdot SMB + h \cdot HML,$$

Following a similar approach Carhart [40] suggested a fourth momentum factor defined as the difference in returns between a portfolio of winners and losers stocks during the previous year. The constant term α in these multi-factor regression models can be used as a measure of a fund's performance, while controlling for the factors included in the models.

5.4.2 Other Fund Assessment Systems and Methodologies

Despite the widespread use of risk-return performance measures in academic research as well as by professional portfolio analysts, their understanding and use by individual investors is a difficult task. In response to the growing need for easy to use investment information, rating agencies provide rating assessment for a wide range of mutual funds. A typical example is the star rating system of Morningstar, which classifies funds into five performance categories (star ratings) according to their risk-adjusted performance relative to predefined peer groups which consist of funds of the same investment style. The top 10 % performing funds within a peer group are assigned with five stars (best performers), the next 22.5 % receive four stars, the next

35 % receive three stars, the next 22.5 % receive two stars, while the bottom 10 % corresponds to the class of worst performing funds (one star).

Nevertheless, such rating systems do not provide a universal coverage of all funds in all countries. Furthermore, apart for the risk-return attributes of the funds, other issues, such as management fees, investment expenses, and transaction costs, are also highly relevant. Operations research methods can be helpful in this context. For instance, several studies have used frontier methods such as data envelopment analysis [150, 156, 173, 199], stochastic frontier analysis [6, 119], and free disposal hull [141], to measure the efficiency of mutual funds in an input-output production framework. In this context, the outputs involve return measures, whereas the inputs include attributes related to risk and the operating characteristics of the funds (e.g., expense ratio, loads, turnover, size, etc.).

5.4.3 Appraisal and Fund Portfolio Optimization with Multiple Criteria

MCDA methods have been used in the context of mutual fund investments for both fund appraisal and the construction of fund portfolios. In the former case, fund ranking and classification models are applicable. Pendaraki et al. [196] used the UTADIS method to build fund selection models for Greek mutual funds. The model classified the funds in terms of their future excess return over a market index in two categories: (a) the "winners", including funds with positive excess return, and (b) the "losers" involving funds with negative excess return. Two settings were considered for the definition of these categories, one being more conservative and the second one being more aggressive. Under the conservative scenario a fund i was classified in the high performance (winners) categories if $r_i > 1.05r_M$, where r_i is the annual return of the fund during the next year and r_M is the corresponding return of the market. For the aggressive scenario the classification rule was modified as $r_i > 1.1r_M$.

The models combined several performance criteria, including: standard deviation of past returns, change in net asset value, geometric mean of past returns, the Sharpe index, systematic risk, Jensen's α, Henriksson-Metron α and γ coefficients. The performance measures where selected on the basis of the literature in the field and their explanatory power. In order to test the validity and predictive performance of the models, a leave-one-out cross-validation (LOO-CV) procedure was employed. LOO-CV is a re-sampling technique, under which M training and test runs are performed using a given data set of M mutual funds. In each run i, a model is constructed from the original data set, excluding mutual fund i. The model is then applied to classify the excluded fund. The expected predictive performance of the model can then be estimated from the results for all mutual funds left out of the training process during this iterative procedure.

The results obtained from the MCDA approach under this evaluation procedure were compared against other popular parametric and non-parametric classification

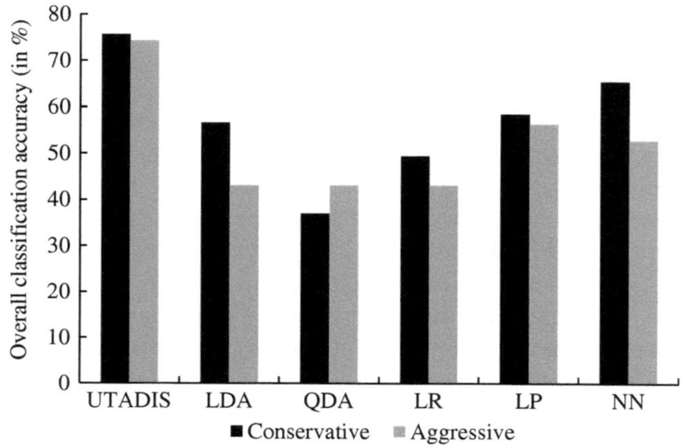

Fig. 5.4 Comparison of classification accuracy in mutual fund classification according to the results in [196]

techniques, including linear and quadratic discriminant analysis, logistic regression, a linear programming formulation, and a nearest neighbor algorithm. Figure 5.4 illustrates the obtained results (overall classification accuracy). It is clearly evident that the predictive power of the multicriteria model was considerably higher than the other techniques.

In a second stage, the construction of a portfolio was considered, consisting of funds selected according to the MCDA fund classification models. The portfolio optimization process was based on nine criteria related to the funds' risk-return characteristics. The optimization was performed through a non-linear goal programming model, solved with different priorities given to the objectives. The resulting portfolios (under both the conservative and aggressive scenarios) were found to outperform the market benchmark in an out of sample test (i.e., future time period) in terms of their returns.

In a different setting, Babalos et al. [12] used a simulation-based approach to assess the performance of US mutual funds. Simulation techniques have been used in MCDA for addressing the uncertainty with regard to the data and the preferential parameters of decision models, thus facilitating the formulation of robust conclusions under a wide range of different performance evaluation scenarios [151, 239]. In a context of mutual fund evaluation, such an approach enables the consideration of different settings and hypotheses with respect to the investment policy and risk attitude of fund managers and investors. Assume a fund evaluation model $F(\mathbf{x}; \boldsymbol{\alpha})$, where $\boldsymbol{\alpha}$ is vector of parameters of the evaluation model, defined on the basis of the investment policy and objectives of the decision maker. Under a simulation approach S acceptable scenarios are considered, each corresponding to different vectors $\boldsymbol{\alpha}_1, \ldots, \boldsymbol{\alpha}_S$. The funds are evaluated under each scenario and the results are used to form a holistic evaluation and obtain different statistics of the performance of the funds. For instance, assuming that a decision model $F(\mathbf{x}; \boldsymbol{\alpha})$ provides an assessment (global score) and ranking of a set of M funds, such that the higher the global score of a fund, the higher is its overall performance, a holistic evaluation can be constructed through aggregation procedures such as:

- Holistic acceptability index [151]:

$$H(\mathbf{x}_i) = \sum_{r=1}^{M} \left(\sum_{\ell=r}^{M} \frac{1}{\ell} \Big/ \sum_{\ell=1}^{M} \frac{1}{\ell} \right) p_{ir}$$

  where p_{ir} is the percentage of scenarios under which fund i is ranked in position r ($r = 1$ corresponds to the best performing fund and $r = M$ to the worst performing one).
- Borda count:

$$B(\mathbf{x}_i) = \frac{1}{M-1} \sum_{r=1}^{M} (M - r) p_{ir}$$

- Average performance score:

$$\bar{F}(\mathbf{x}_i) = \frac{1}{S} \sum_{s=1}^{S} F(\mathbf{x}_i; \boldsymbol{\alpha}_s)$$

Babalos et al. [12] implemented this approach using an additive value function as the evaluation model. The multicriteria evaluation model was applied to a panel data set of 485 fund-year observations involving US mutual funds over the period 2000–2009. The evaluation criteria included the following performance attributes:

- Total expenses ratio, defined as the total costs charged by a fund (management fees and other operational and administrative costs) over its average annual net assets.
- Front-end loads, which involve the commission paid by an investor upon the initial investment in a fund.
- Annualized standard deviation of the returns, which defines the risk of the investment.
- Deviation of a fund's return from the median return, which provides an estimate of a fund's return relative to its peer group in the same time-period (a year).

Apart for these criteria, two risk-adjusted performance measures were also considered, namely Jensen's α and Carhart's α. An examination of the aggregation of the simulation results with the three aforementioned approaches (holistic acceptability, Borda count, average score) revealed only minor differences in the evaluation of the funds. As far as the relative importance of the selected performance criteria is concerned, the results verified the significance of the two risk-adjusted measures. The risk and return criteria were also found important, whereas the two attributes related to the costs and fees of the funds (expenses ratio and front-end loads) were the least important performance measures. The risk-adjusted attributes and the return criterion were also found to be highly correlated with the changes in the evaluation of the funds over time (i.e., funds that managed to improve their performance versus funds whose performance decreased).

5.5 Index Tracking

Portfolio management strategies can be classified as active or passive. Active strategies focus on frequent transactions that seek to re-balance the composition of a portfolio in order to maximize return. Such strategies exploit imperfections and inefficiencies in the markets, which (when they exist) may create short-term profit opportunities for investors. On the other hand, passive strategies assume that markets are efficient and thus, they focus on constructing and holding well-diversified portfolios, which will be profitable in the long-run.

A popular passive management approach involves the construction of portfolios that closely match a pre-defined index. By tracking an index, the unsystematic component of risk is diversified away and transaction costs due to frequent re-balances of the portfolio are minimized. Index tracking strategies are particularly important for fund managers as many mutual funds and exchange traded funds are based on tracking market indices.

Index tracking can be implemented following either a full or a partial replication approach. Full replication is based on portfolios consisting of all stocks in an index. Alternatively, portfolios of only a limited number of stocks can be considered, thus lead to partial replication. Partial replication is easier to implement because it is based on portfolios with a small number of assets, thus leading to lower management and transaction costs. Furthermore, it would also be possible to consider an index replication strategy with stocks not actually belonging to the replicating index (e.g., from a different stock market) or even with different asset classes [4].

Formally, assume a set of M assets available for constructing a portfolio of at most U assets that replicates the returns of a given index as closely as possible. In particular, denoting by $\mathbf{r}_I = (r_{I1}, \ldots, r_{IT})$ the returns of an index I over T time periods, the following mean-squared tracking error measure can be defined:

$$MSE = \frac{1}{T}\sum_{t=1}^{T}(r_{pt} - r_{It})^2 = \frac{1}{T}(\mathbf{Rx} - \mathbf{r}_I)^\top(\mathbf{Rx} - \mathbf{r}_I)$$

Under this error measure, the portfolio construction problem is a variant of the mean-variance model (5.1):

$$
\begin{aligned}
\min \quad & (\mathbf{Rx} - \mathbf{r}_I)^\top (\mathbf{Rx} - \mathbf{r}_I) \\
\text{s.t.} : \; & \bar{\mathbf{r}}^\top \mathbf{x} \geq R \\
& \mathbf{aq} \leq \mathbf{x} \leq \mathbf{bq} \\
& \mathbf{1}^\top \mathbf{q} \leq U \\
& \mathbf{1}^\top \mathbf{x} = 1 \\
& \mathbf{q} \in 0, 1
\end{aligned}
$$

Similarly, to the case of portfolio optimization, alternative measures of tracking error can be introduced, based on absolute deviations, CVaR measures, and other downside deviation criteria. Gaivoronski et al. [97] discussed several such tracking error measures (both static and dynamic) and conducted extensive computational comparisons, concluding that "there is no unanimity as to what the best measure of tracking error is". Therefore, the arguments developed earlier for portfolio management also apply in the context of index tracking.

From a computational perspective, it should be noted that portfolio construction with constraints on the number of assets in the portfolio is a computationally difficult problem, particularly when the number of available assets is large. Evolutionary methods and heuristics constitute good alternatives in such cases [44].

5.6 Decision Support Systems

The implementation of analytic techniques into decision support systems (DSSs) greatly facilitates the adoption of such techniques in practice. The area of portfolio management is very suitable for the design and development of powerful DSSs, as investors and analysts face enormous real-time data, which cannot be handled without computer-aided systems designed to integrate data with analytic methods.

DSSs incorporating multicriteria decision aid methods in their structure are known as multicriteria DSSs, and they have found several applications in the field of finance (for an overview see, [267]), including portfolio management. Multicriteria DSSs for portfolio selection combine data management and analysis capabilities, with models from portfolio theory, and MCDA methods for asset screening and capital allocation.

An example of a multicriteria DSS for equity portfolio management is the INVESTOR system [268]. The main characteristic of the system is the combination of portfolio theory models, multivariate statistical methods, and multicriteria decision aid techniques for stock evaluation and portfolio construction. The structure of the system is presented in Fig. 5.5.

The system combines four main types of information:

- financial data drawn from the financial statements of the firms,
- stock market data, such as stock prices, trading volumes, dividend yields, and valuation ratios (price/earnings per share, price/book value),

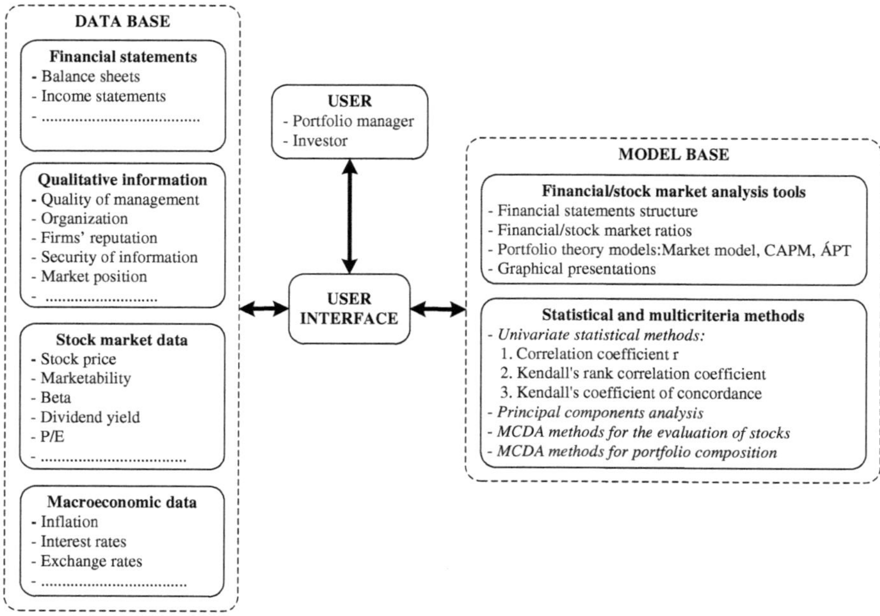

Fig. 5.5 Structure of the INVESTOR system

- qualitative information regarding the management of the firms, their organization, their reputation in the market, their position in their business sector, the innovation and know-how levels, etc.,
- macroeconomic factors such as inflation, interest rates, and exchange rates.

On the basis of these data, the INVESTOR system enables users to perform analyses with well-known portfolio models, such as the CAPM and the APT. With the CAPM, the investor/analyst can produce estimates for the expected stock returns of the firms in relation to their systematic risk and the risk-free rate, whereas APT allows the consideration of additional factors in a multidimensional context.

Furthermore, the system incorporates data analysis techniques, such as rank correlation statistics and principal components analysis that enable the user to analyze the characteristics of the firms, their risk-return patterns, and identify the main factors that describe their performance.

On the decision modeling side, the INVESTOR system uses MCDA techniques for both asset selection and portfolio optimization. The former is performed with two preference disaggregation techniques, which allow the construction of decision models that rank and classify the available stocks into performance groups. The investor can calibrate the models through the system to suit his/her investment policy and objectives by providing (interactively) examples of his/her own assessment of the prospects of the stocks. A similar approach for constructed stock evaluation and selection models was also employed in the DSS proposed in [213]. For the portfolio

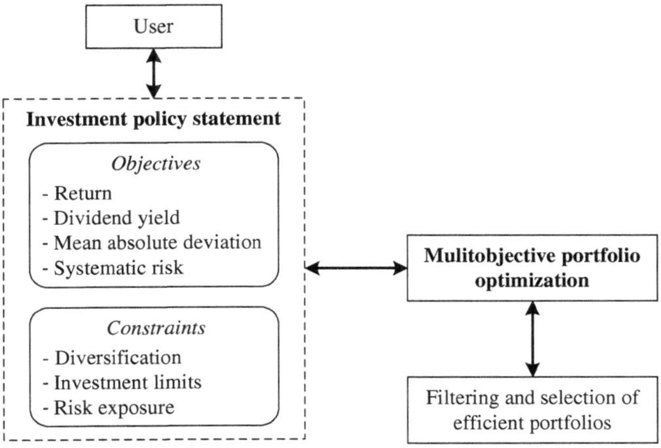

Fig. 5.6 The methodological framework in the IPSSIS system

construction process, the system uses a novel goal programming formulation that penalizes portfolios deviating significantly from the ideal values of the objectives, combined with post-optimality analysis methods to consider the robustness of the results.

Other multicriteria DSSs for portfolio management have focused on specific stages of the investment process. For instance, Xidonas et al. [260] presented the IPSSIS, which focuses on the portfolio optimization stage (Fig. 5.6). The system enables the user to specify his/her investment policy by defining proper objectives and constraints. The set of objectives included in the system involves return, dividend yield, and risk. The risk component is taken into consideration through the beta coefficient for assessing systematic risk and the mean absolute deviation as a measure for unsystematic risk. The user can also define several types of constraints to ensure that the constructed portfolios are well diversified. Among others, these constraints involve the number of assets in the portfolio, the capital invested in specific stocks, business sectors, and capitalization categories, as well as constraints involving stocks with particular risk characteristics (e.g., stocks with low systematic risk). The optimization process is implemented through the augmented ε-constraint method [174], which constructs the complete set of efficient portfolios through an iterative process. Finally, the system incorporates a module that facilitates the selection of the best portfolio according to the investment policy preferences of the user. This is particularly useful as, practically, an infinite number of efficient portfolios can be constructed. To address this issue the system uses an interactive filtering algorithm that enables the user to select a limited set of portfolios according to their performance on the selected performance measures (objectives).

Chapter 6
Other Applications of Multicriteria Analysis in Finance

Abstract This chapter illustrates the contributions of MCDA in other areas of financial decision making. First, the investment appraisal process is considered followed by country risk analysis. For the latter, an illustrative application is presented demonstrating preference disaggregation methods that can be used to construct country risk classification models.

Keywords Investment appraisal · Project selection · Country risk analysis

6.1 Investment Appraisal

Decisions on the choice of investment projects often have a strategic character as they span over a large time period and they require considerable resources. The investment decision process consists of four main stages: perception, formulation, evaluation and choice. The financial theory is mostly involved with the evaluation and choice stages, through the introduction of investment appraisal criteria such as the net present value, the internal rate of return, and the payback method. Such criteria are aggregated through empirical approaches resulting to a ranking of a set of investment projects on the basis of their attractiveness or to an acceptance/rejection decision in the case of a single project.

However, there are a number of issues with the above process. First, the analysis is restricted to the evaluation of future cash flows on the basis of a predefined discount rate. Secondly, there is no formal framework for analyzing the discrepancies in the results of different investment appraisal criteria. In a realistic setting, the investment analysis is much more involved than a simple discounting of future financial outcomes. Furthermore, the high uncertainties involved with the outcomes of an investment project cannot always be adequately described in probabilistic terms, especially in cases of strategic investments for which similar past instances or historical data are not available.

M. Doumpos and C. Zopounidis, *Multicriteria Analysis in Finance*,
SpringerBriefs in Operations Research, DOI: 10.1007/978-3-319-05864-1_6,

Instead, a comprehensive investment appraisal process requires the careful consideration of possible options (investment projects), the specification of the goals and objectives of the investments, the identification of their consequences and risks, as well as the formulation of the evaluation results. The multicriteria paradigm introduces such a holistic view of the investment selection process, supporting all of its stages. Montibeller et al. [180] analyzed the contributions of MCDA in the problem structuring phase, in the context of project portfolio selection. Concerning the stages of evaluation and choice, MCDA offers a methodological framework much more realistic than the one based solely on financial criteria, which make assumptions that are often not met in practice. For instance, Götze et al. [103] note that investment appraisal based on the net present value, assumes among others that:

1. a single performance measure is adequate,
2. the economic life of the investment is known,
3. the investment appraisal process is separated from other relevant decisions regarding the financing of the project and its operation,
4. the cash flows are known.

In fact, the financial outcomes of the project and the associated risks depend on a number of factors, which are often difficult to quantify. For instance one can mention the strategic benefits of the investment, its relation to the organization strategy of the firm, technical aspects of the investment, operational risk factors related to the implementation of the investment, regulatory and legal issues, etc. Recently new trends have also emerged with regard to socially responsible investments, thus adding ethical, social, and environmental criteria in the analysis.

The multidimensional nature of the investment appraisal process is further highlighted by the multiple objectives that managers seek to achieve through the implementation of an investment project. Bhaskar and McNamee [26] presented empirical results from large companies from the United Kingdom, showing that 96 % of the companies consider more than one objective during the investment selection process (with the most common number of objectives being eight). In most cases, profitability was found to be given top priority, followed by company growth, risk, liquidity, flexibility, etc.

In a venture capital investment context, empirical survey studies have presented extensive empirical results from survey studies conducted among US, UK, and European venture capital firms, in order to identify the criteria that they consider in their investment process [62, 109, 159, 184]. The results demonstrate that such investment decisions are driven by a diverse set of qualitative and quantitative factors, involving among others:

- the qualities and experience of the management team of the firms,
- the experience and personality of the entrepreneurs,
- product-market criteria,
- the financial characteristics of the investments,
- the lending guidelines followed by the venture capital firms, etc.

The aggregation of such a diverse set of decision criteria in an ad-hoc manner, without a solid, structured, and sound framework underlying the characteristics of the evaluation process can easily lead to flawed and unexpected results. For instance, Keeney [138] analyzes 12 common mistakes in making value trade-offs, which are also relevant in other evaluation contexts. Among the most generally applicable ones, we can mention the following:

- not understanding the decision context,
- not having measures for consequences (i.e., criteria),
- using inadequate measures,
- not knowing what the measures represent,
- replacing fundamental objectives with alternative proxies,
- focusing on calculating "correct" trade-offs,
- using screening criteria imposing value judgments,
- failure to use consistency checks.

The MCDA paradigm provides investors and managers with a systematic approach to handle such issues, thus enabling the consideration of the investment appraisal process in a realistic and flexible multicriteria context. Among others, MCDA techniques, which are applicable in investment appraisal are involved with issues such as:

1. facilitating the managers in specifying a solid and transparent structure of the investment selection process,
2. analyzing the trade-offs among the investment selection criteria and measuring their relative importance,
3. aggregating multiple appraisal measures of diverse nature (qualitative, quantitative, deterministic, stochastic, fuzzy, etc.) into global investment selection indices,
4. exploring the uncertainties involved in the selection process, through systematic sensitivity and robustness analyses.

Table 6.1 reports some recent studies using MCDA approaches for investment appraisal in different contexts.

6.2 Country Risk Analysis

6.2.1 The Context of Country Risk Assessment

The oil crises of the 1970s and the resulting worldwide economic turmoil were the first post-war events that highlighted the importance of a global risk factor for sustainable socio-economic development as well as for the operation of firms worldwide. More recent events, such as the crises in Southeast Asia (1997), South America (2002), as well as the global credit crisis of 2007–2008 and the subsequent European sovereign debt crisis are clear examples that demonstrate the relevance of country risk for financial decision making.

Table 6.1 Some recent studies on investment appraisal under multiple criteria

Information and communication technologies [5]
Army modernization [41]
Transport [55]
International project portfolios [112]
Cash flow modeling [132]
Capital budgeting under fuzziness and uncertainty [149]
Transport [158]
Energy systems [192]
Shipping [206]
Wind farm site selection [108]
Product design [253]

Country risk has many facets, which arise from the different perspectives that financial decision makers view the economic and financial development of a country and the difficulties that it faces. From an economic perspective, country risk can be defined as the probability that a country will fail to generate enough foreign exchange to pay its obligations toward its foreign creditors [50]. This economic point of view, however, is focused on the capacity of a country to service its debt. Socio-economic factors are also highly relevant, as they represent the willingness of a country to service its debt. In that regard, country risk can be defined in broader context as the potential economic and financial losses due to the difficulties raised from the macro-economic and/or political environment of a country [38]. Such a definition covers not only the losses for the creditors of a country (financial institutions, organizations, other countries, etc.), but also losses that any corporate entity and institutional or private investor may experience for investments undertaken in a country. For instance, Claude et al. [45] analyzed the relevance and applications of country risk analysis to the portfolio management process, including equity and fixed income portfolios. On the other hand, from the perspective of corporate financial investments, macro and micro risks can be further identified [115, 243]. Macro (sociopolitical) risks arise from dramatic events such as wars, sectarian conflicts, revolutions, etc., as well as less dramatic events such as the country-wide imposition of price controls, tax increases or surcharges, etc. Micro risks, on the other hand, concern circumstances involving industry, firm or project-specific cancellation of import and export licenses, discriminatory taxes, etc.

As an example of the issues involved in country risk analysis, one may consider the diversity of the factors examined by the three main credit rating agencies (Moody's, Standard and Poor's, Fitch), which include [94]:

- Macro-economic conditions and growth factors related to the scale of the economy in a country, its competitiveness, its ability to achieve sustainable growth, and the effectiveness of monetary policies.

- Public finance factors describing the ability of a government's revenue-raising efficiency, its effectiveness in handling expenditures, managing its assets, and obtaining foreign currency.
- Debt factors related to the level, structure, and dynamics of public debt.
- Financial sector attributes that focus on the strength of a country's financial sector, its effectiveness, and the quality of its supervision.
- External finances related to the balance of payments, foreign exchange reserves adequacy, and the structure of the current account.
- Exchange rate regimes and their compatibility to a country's monetary goals.
- Political factors, including geopolitical risk, policy transparency, international relations, public security, as well as the stability and legitimacy of political regime in a country.
- Structural and institutional factors covering issues such as corruption, transparency, institutional independence, the efficiency of the public sector, the strength of the business environment, and the level of innovation.
- Other factors related to the labor market, the openness of the economy, as well as risks from natural disasters.

The first attempts to establish country risk assessments were mainly based on checklist systems focused on economic variables. However, this approach has been proven to be insufficient mainly due to its inability to establish a sound methodological framework for the selection and weighting of the variables. To address this issue, several statistical techniques have been used, mainly oriented towards building models for analyzing and predicting debt reschedulings (for an overview, see [144]) and the country risk ratings issued by rating agencies and international organizations. An overview of international practices in country risk ratings and their primary dimensions can be found in Claude at el. [45], whereas a recent report by the International Monetary Fund focuses on the ratings issued by the three major rating agencies (Standard and Poor's, Moody's, Fitch) and analyzes their role in the recent global crisis as well as their accuracy and information value [94] (Chap. 3).

6.2.2 Multicriteria Approaches to Country Risk Analysis

The MCDA methodologies have been used for country risk assessment to develop models that rank or classify countries into risk groups. Tang and Espinal [237] developed a multiattribute model to assess country risk, both on a short and medium-long term basis. The model considered 14 risk criteria related to the countries' external repayment capability, their liquidity, per capital income and population growth, as well as purchasing power risk. The selection and weighting of the criteria was based on the Delphi method. The model was applied to a sample of 30 developed and developing countries. The results showed that the most significant country risk indicator both for short and medium-long terms was the external repayment capability of a country. The ranking of the countries according to the multicriteria model was found to be consistent with the evaluations of two international banks.

Oral et al. [189] proposed a generalized logistic regression model to assess country risk. The parameters of the model were estimated through a mathematical programming formulation controlling for the geopolitical economic characteristics of the countries. The model reproduced the country risk rating scores of Institutional Investor and it was applied to a sample of 70 countries for the years 1982 and 1987. A comparison with logistic regression and regression trees indicated the superiority of the new method over statistical models. Regarding the importance of country risk indicators, the three models provided similar results, highlighting the importance of indicators such as debt/exports, gross national product (GNP) per capita, and investments/GNP.

Cosset et al. [50] applied a preference disaggregation methodology for the development of a country risk ranking model, based on the UTASTAR multicriteria method [221]. Using a sample of 22 reference countries, an additive value model was interactively developed, which consistently represented the preferences of a decision maker. The most important determinants of sovereign creditworthiness were found to be the GNP per capita ratio, propensity to invest, as well as the current account balance to GNP ratio.

6.2.2.1 An Illustration for Country Risk Classification

Except for the above studies that focused on multicriteria models for ranking countries, classification approaches have also been used. Multicriteria classification techniques are particularly well-suited to country risk assessment as they enable the construction of risk rating models that assign countries into predefined risk categories, in accordance with rating systems commonly used by investors, policy makers, and financial risk analysts.

Following such an approach Doumpos et al. [66] used the MHDIS method (Multi-group Hierarchical DIScrimination [266]) for the construction of a classification model. Similarly to the UTADIS method (see Sect. 4.4), the MHDIS method also employs a value function modeling approach. However, often alternatives (e.g., countries) belonging into different performance categories may have very different characteristics, thus making a single scoring model unable to fully describe the data and discriminate the categories. To address this issue, the MHDIS method leads to the construction of multiple value functions. For instance, assume a country risk classification problem in which countries are grouped in N ordered risk categories $C_1, \ldots, C_N$, defined such that C_1 is the low risk group and C_N the high risk one. The modeling approach of the MHDIS method is based on $N-1$ pairs of value functions $\{V_\ell(\mathbf{x}), V_{\sim\ell}(\mathbf{x})\}$, $\ell = 1, \ldots, N-1$, where $V_\ell(\mathbf{x})$ is the value function corresponding to risk category C_ℓ and $V_{\sim\ell}(\mathbf{x})$ describes countries in higher risk classes $C_{\ell+1}, \ldots, C_N$. The two evaluation functions are parameterized by different trade-offs and marginal value functions, each representing the characteristics of countries in class C_ℓ versus countries in categories $C_{\ell+1}, \ldots, C_N$. Under this setting a country i is classified to the risk category with the lowest index ℓ^*, such that $V_{\ell^*}(\mathbf{x}_i) \geq V_{\sim\ell^*}(\mathbf{x}_i)$.

The value functions have a piecewise linear additive form similar to the one described in Sect. 4.4. They are constructed using a preference disaggregation approach that combines three optimization models. The first model is a linear program that minimizes the weighted sum of all absolute errors for the countries in a reference (training) sample, on the basis of the above classification rule:

$$\begin{array}{lll}
\min & \sum_{\ell=1}^{N} \frac{1}{M_\ell} \sum_{\mathbf{x}_i \in C_\ell} (\varepsilon_{\ell i}^+ + \varepsilon_{\ell i}^-) & \\
\text{s.t.} & V_\ell(\mathbf{x}_i) - V_{\sim\ell}(\mathbf{x}_i) + \varepsilon_{\ell i}^+ \geq \delta, & \forall \mathbf{x}_i \in C_\ell \\
& V_\ell(\mathbf{x}_i) - V_{\sim\ell}(\mathbf{x}_i) - \varepsilon_{\ell i}^- \leq -\delta, & \forall \mathbf{x}_i \in \{C_{\ell+1}, \ldots, C_N\} \\
& \varepsilon_{\ell i}^+, \varepsilon_{\ell i}^- \geq 0 &
\end{array}$$

where δ is a user-defined small positive constant. At a second stage, the classification results from the model derived from the above formulation, are calibrated to reduce the number of misclassifications. In particular, let *MIS* denote the countries misclassified according to the set of additive value functions resulting from the above linear program. The objective of the second stage is to minimize the number of these cases, while retaining all the correct assignments for the other countries (set *COR* of correctly classified countries). This is achieved through the following mixed-integer program:

$$\begin{array}{lll}
\min & \sum_{\ell=1}^{N} \frac{1}{M_\ell} \sum_{\mathbf{x}_i \in C_\ell \cap MIS} (y_{\ell i}^+ + y_{\ell i}^-) & \\
\text{s.t.} & V_\ell(\mathbf{x}_i) - V_{\sim\ell}(\mathbf{x}_i) \geq \delta, & \forall \mathbf{x}_j \in C_\ell \cap COR \\
& V_\ell(\mathbf{x}_i) - V_{\sim\ell}(\mathbf{x}_i) \leq -\delta, & \forall \mathbf{x}_i \in \{C_{\ell+1}, \ldots, C_N\} \cap COR \\
& V_\ell(\mathbf{x}_i) - V_{\sim\ell}(\mathbf{x}_i) + y_{\ell i}^+ \geq \delta & \forall \mathbf{x}_i \in C_\ell \cap MIS \\
& V_\ell(\mathbf{x}_i) - V_{\sim\ell}(\mathbf{x}_i) + y_{\ell i}^- \leq -\delta, & \forall \mathbf{x}_i \in \{C_{\ell+1}, \ldots, C_N\} \cap MIS \\
& y_{\ell i}^+, y_{\ell i}^- \in \{0, 1\} &
\end{array}$$

The first two constraints ensure that all correct classifications achieved at the first stage are retained, whereas the following two constraints are only used for misclassified countries. The binary error variables y^+ and y^- indicate whether a country is misclassified or not (in the former case they equal one, otherwise they are zero).

The result of the above mixed-integer formulation provides the best discrimination of the countries in the risk categories, in term of the number of misclassifications. The last stage of the model fitting process involves a final calibration in order to achieve robust results. For a country i correctly classified in risk category C_ℓ, the pair of value functions $\{V_\ell(\mathbf{x}), V_{\sim\ell}(\mathbf{x})\}$ provides a robust result if the difference $V_\ell(\mathbf{x}_i) - V_{\sim\ell}(\mathbf{x}_i)$ is maximized. Similarly, the pair of value functions $\{V_\ell(\mathbf{x}), V_{\sim\ell}(\mathbf{x})\}$ provides a robust result for a country i correctly classified in risk categories $\{C_{\ell+1}, \ldots, C_N\}$ if the difference $V_{\sim\ell}(\mathbf{x}_i) - V_\ell(\mathbf{x}_i)$ is maximized. In that regard, denoting by COR' and MIS' the set of countries classified, respectively correctly and incorrectly, by the value functions developed through the above mixed-integer programming model, the last stage involves the solution of the following linear program:

$$
\begin{array}{lll}
\max & d & \\
\text{s.t.} & V_\ell(\mathbf{x}_i) - V_{\sim\ell}(\mathbf{x}_i) - d \geq 0, & \forall\, \mathbf{x}_i \in C_\ell \cap COR' \\
& V_\ell(\mathbf{x}_i) - V_{\sim\ell}(\mathbf{x}_i) + d \leq 0, & \forall\, \mathbf{x}_i \in \{C_{\ell+1}, \ldots, C_N\} \cap COR' \\
& V_\ell(\mathbf{x}_i) - V_{\sim\ell}(\mathbf{x}_i) \leq 0 & \forall\, \mathbf{x}_i \in C_\ell \cap MIS' \\
& V_\ell(\mathbf{x}_i) - V_{\sim\ell}(\mathbf{x}_i) \geq 0, & \forall\, \mathbf{x}_i \in \{C_{\ell+1}, \ldots, C_N\} > \cap\, MIS' \\
& d \geq 0 &
\end{array}
$$

The first pair of constraints involve only the correctly classified countries. In these constraints, d represents the minimum absolute difference between the global values of each country according to the two value functions, which must be maximized in order to ensure that the obtained results are robust. On the other hand, the second pair of constraints involves the misclassified countries, and it is used to ensure that they will be retained as misclassified. The set of value functions resulting from the linear program can then be employed to classify any country outside the reference sample.

Following this multicriteria approach, Doumpos et al. [66] used a sample of 161 countries over the period 1996–2000. The countries were classified into four groups according to their income classification as defined by the World Bank:

1. High-income economies (class C_1), including 31 countries.
2. Upper-middle income economies (class C_2), including 30 countries.
3. Lower-middle income economies (class C_3), including 44 countries.
4. Low-income economies (class C_4), including 56 countries.

It should be noted, however, that such a classification is only a rough proxy of country risk, as it is focused on the countries' wealth and does not explicitly consider their economic ability to service their debt, or other socio-economic factors that contribute to country risk as explained earlier.

With this limitation in mind, the evaluation of the countries was performed through 12 country risk indicators selected on the basis of the literature on country risk assessment, and their discriminating power in the context of the specific data. The selected indicators and their trade-offs as estimated through the MHDIS method are reported in Table 6.2. The differences in the obtained results are indicative of the diverse characteristics of the four performance categories of countries in the sample. For instance, countries in the high-income group (function V_1) are characterized by high current account balance, high investments (foreign direct investments and capital formation), and low debt service payments.

Given that the countries are classified in four categories, the classification model consists of three pairs of additive value functions, according to which a country i is classified as follows:

$$
\begin{array}{l}
\text{If } V_1(\mathbf{x}_i) > V_{\sim 1}(\mathbf{x}_i), \text{ then } \mathbf{x}_i \in C_1 \\
\quad \text{else if } V_2(\mathbf{x}_i) > V_{\sim 2}(\mathbf{x}_i), \text{ then } \mathbf{x}_i \in C_2 \\
\quad\quad \text{else if } V_3(\mathbf{x}_i) > V_{\sim 3}(\mathbf{x}_i), \text{ then } \mathbf{x}_i \in C_3 \\
\quad\quad \text{else } \mathbf{x}_i \in C_4
\end{array}
$$

Table 6.2 Trade-offs of the country risk indicators in the the MHDIS classification model (in %)

Criteria	V_1	$V_{\sim 1}$	V_2	$V_{\sim 2}$	V_3	$V_{\sim 3}$
Current account balance/GDP	16.55	0.78	6.57	28.06	22.36	2.19
Exports of goods and services/GDP	0.80	0.80	0.65	0.65	11.28	8.11
Foreign direct investment/GDP	10.88	0.73	0.59	9.69	0.76	3.11
Gross capital formation/GDP	17.44	0.79	12.58	0.64	3.49	0.50
Inflation	4.21	13.06	2.54	5.66	5.74	0.51
Infant mortality rate	0.77	47.95	26.25	19.84	11.33	20.49
Short-term debt/Total external debt	8.74	0.61	0.61	0.61	0.49	0.49
Total debt service/Exports of goods and services	15.38	2.67	7.75	1.79	0.49	11.71
Total debt service/Gross international reserves	0.60	0.60	6.93	0.60	12.03	0.49
Net domestic credit/GDP	6.60	24.20	33.90	4.29	6.99	29.10
Total external debt/GDP	0.42	6.73	1.20	27.80	16.12	16.21
Total debt service/Gross international reserves	17.61	1.09	0.43	0.38	8.91	7.10

Table 6.3 Classification accuracies (in %)

	2000	1999	1998	1997	1996
MHDIS	94.25	83.67	83.43	81.81	81.52
UTADIS	83.96	81.43	79.76	80.57	79.23
Rough sets	100.00	83.37	74.48	77.72	73.80
Neural networks	92.64	84.32	80.53	79.09	74.79
Discriminant analysis	77.33	76.46	73.93	76.47	76.69
Ordinal logistic regression	72.39	76.17	69.19	67.41	66.58

Table 6.3 presents the overall classification accuracy results for the MHDIS method as well as for UTADIS, and four other popular machine learning and statistical techniques (rough sets, neural networks, discriminant analysis, ordinal logistic regression). The 2,000 data were used for fitting the models (i.e., training data), whereas the previous years were used for back-testing the models in order to assess their discriminating power. The results show that the two MCDA methods provide the best results in this back-testing comparison. The model developed with the MHDIS method has an accuracy rate consistently higher than 80 % in all years. The models constructed with rough sets and neural networks outperform the two statistical methods, but their performance is not robust over time. Even though these models perform exceptionally well in the training data for year 2000, their performance in the back-tests decreases considerably reaching 73–74 % in 1996.

Chapter 7
Conclusions and Future Perspectives

Similarly to models in other sciences, financial models are nothing more than mathematical representations of complex financial phenomena, based on assumptions, hypotheses, and simplifications that facilitate the model building and solution process. In a highly volatile global environment, building accurate models becomes a very challenging task. Relaxing the set of assumptions and simplifications leads to more realistic but also more involved models.

The multicriteria paradigm introduces a decision-theoretic approach to financial decision making, based on the simple fact that decisions are taken by actual decision makers instead of models. In this context, each particular decision situation requires the consideration of domain knowledge from the theory and practice of finance, which derives from normative and descriptive financial models, but also a prescriptive and constructive approach that will support the financial decision maker in evaluating and designing proper ways of action suitable for the problem at hand. MCDA contributes towards this direction, providing analytic techniques for supporting all stages of the decision process with emphasis on incorporating all relevant decision criteria (qualitative and quantitative) in the analysis.

The introduction of multiple criteria contributes in relaxing the assumptions that increase model risk, facilitates the learning process of financial decision makers, and ultimately leads to more informed decisions. The techniques and methodologies available in the field of MCDA introduce a systematic and formal approach for addressing the conflicts arising from the consideration of multiple points of view, criteria, and objectives, thus avoiding empirical ad-hoc solutions that not well-grounded on a systematic treatment of their assumptions, consequences, and validity.

Nevertheless, the rapid transformation of the global financial and business environment, along with the new technological developments and innovations, provide new opportunities but also raise important challenges for the use of analytic methods in financial decision support. This also involves the field of MCDA and its uses in this field. In that regard, one can mention a number of important issues, such as:

M. Doumpos and C. Zopounidis, *Multicriteria Analysis in Finance*,
SpringerBriefs in Operations Research, DOI: 10.1007/978-3-319-05864-1_7,

- Strengthening the connections and synergies with the latest developments in financial risk management, behavioral finance, and financial economics, using updated data, research findings, exploring new application areas, and covering new financial instruments and services. This is fundamental for the adoption of MCDA models by finance researchers and professionals. Given that existing MCDA techniques are "general purpose" decision support tools, the possibility of introducing specific concepts, theories, and practices from the field of finance, should be explored to ensure that MCDA models best match the special features of the financial environment.
- Introduction of systematic ex ante and ex post validation procedures for multicriteria models under financial performance measures in accordance with the requirements imposed by the regulatory environment. The realism and appealing features of multicriteria systems for financial problems, are not enough for their practical adoption in finance. The financial regulatory framework has become much stricter as far as it concerns the validity and effectiveness of the models used to support financial decisions. Therefore, MCDA models and techniques need to be further validated through comparative computational results and rigorous model validation tests, based not only on decision-theoretic criteria, but also using measures that are relevant from a financial perspective. Such an approach will highlight not only the methodological contributions of MCDA, but also the added value that it brings compared to existing and well-established financial models.
- Introduction of computational improvements that will allow existing models and algorithms to scale up to massive financial data, in a real-time decision support context. Data derived from the global markets and corporate information are of massive size. Therefore, computational analytic techniques for decision support should be able to handle the dimensionality of the data in an efficient manner. Several algorithmic advances in multi-objective optimization allow the analysis of large-scale problems. Many discrete MCDA methods, however, are mainly focused on smaller data. Extending, such techniques to enable the handling of large sets of alternatives is an important issue for their successful use in finance decision support.
- Implementation into decision support systems, taking advantage of new technologies from the fields of information systems and computer science. Computer-aided support through user-friendly systems is of major importance for providing decision aid through the integration of data and analytic methods. Such systems could be either stand alone (such as the ones illustrated in Sect. 5.6) or based on distributed environments such as client-server architectures and web-based technologies, which provide new capabilities for monitoring, retrieving, processing, and analyzing financial and business data.
- Integration of other emerging areas in operations research and computational intelligence (e.g., data mining, evolutionary algorithms, fuzzy systems, and other soft computing technologies), which will further strengthen the applicability of the multicriteria paradigm in financial domains of high dimensionality/complexity, non-linearity, and uncertainty.

References

1. Abdou, H.A., Pointon, J.: Credit scoring, statistical techniques and evaluation criteria: a review of the literature. Intell. Syst. Account. Finance Manage. **18**(2–3), 59–88 (2011)
2. Altman, E., Fargher, N., Kalotay, E.: A simple empirical model of equity-implied probabilities of default. J. Fixed Income **20**(3), 71–85 (2011)
3. Altman, E., Saunders, A.: Credit risk measurement: developments over the last 20 years. J. Bank. Finance **21**(11–12), 1721–1742 (1997)
4. Andriosopoulos, K., Doumpos, M., Papapostolou, N.C., Pouliasis, P.K.: Portfolio optimization and index tracking for the shipping stock and freight markets using evolutionary algorithms. Transp. Res. Part E **52**, 16–34 (2013)
5. Angelou, G.N., Economides, A.A.: A compound real option and ahp methodology for evaluating ICT business alternatives. Telematics Inform. **26**(4), 353–374 (2009)
6. Annaert, J., van den Broeck, J., Vander Vennet, R.: Determinants of mutual fund underperformance: a Bayesian stochastic frontier approach. Eur. J. Oper. Res. **151**(3), 617–632 (2003)
7. Aouni, B., Colapinto, C., La Torre, D.: Financial portfolio management through the goal programming model: current state-of-the-art. Eur. J. Oper. Res. **234**(2), 536–545 (2014)
8. Arrow, K.J., Raynaud, H.: Social Choice and Multicriterion Decision-making. MIT Press, Cambridge (1986)
9. Artzner, P., Delbaen, F., Eber, J.M., Heath, D.: Coherent measures of risk. Math. Finance **9**(3), 203–228 (1999)
10. Azmi, R., Tamiz, M.: A review of goal programming for portfolio selection. In: Jones, D., Tamiz, M., Ries, J. (eds.) New Developments in Multiple Objective and Goal Programming. Lecture Notes in Economics and Mathematical Systems, vol. 638, pp. 15–33. Springer, Berlin (2010)
11. Aznar Bellver, J., Cervelló Royo, R., García García, F.: Spanish savings banks and their future transformation into private capital banks: determining their value by a multicriteria valuation methodology. Eur. J. Econ. Finance Adm. Sci. **35**, 155–164 (2011)
12. Babalos, V., Philippas, N., Doumpos, M., Zopounidis, C.: Mutual funds performance appraisal using stochastic multicriteria acceptability analysis. Appl. Math. Comput. **218**(9), 5693–5703 (2012)
13. Kalyebara, B., Islam, S.M.N.: Corporate Governance, Capital Markets, and Capital Budgeting–An Integrated Approach. Springer, Berlin (2014)
14. Ballestero, E.: Stochastic goal programming: a mean-variance approach. Eur. J. Oper. Res. **131**(3), 476–481 (2001)
15. Ballestero, E., Bravo, M., Perez-Gladish, B., Arenas-Parra, M., Pla-Santamaria, D.: Socially responsible investment: a multicriteria approach to portfolio selection combining ethical and financial objectives. Eur. J. Oper. Res. **216**(2), 487–494 (2012)

M. Doumpos and C. Zopounidis, *Multicriteria Analysis in Finance*,
SpringerBriefs in Operations Research, DOI: 10.1007/978-3-319-05864-1,

16. Ballestero, E., Romero, C.: Portfolio selection: a compromise programming solution. J. Oper. Res. Soc. **47**(11), 1377–1386 (1996)
17. Banker, R.D., Charnes, A., Cooper, W.W.: Some models for the estimation of technical and scale inefficiencies in data envelopment analysis. Manage. Sci. **30**, 1078–1092 (1984)
18. Basel Committee on Banking Supervision: International convergence of capital measurement and capital standards: a revised framework. Bank for International Settlements, Basel, Switzerland (2004)
19. Barberis, N., Thaler, R.: A survey of behavioral finance. In: Constantinides, G.M., Harris, M., Stulz, R.M. (eds.) Handbook of the Economics of Finance, vol. 1, Chap. 18, pp. 1053–1128. Elsevier, Amsterdam (2003)
20. Baron, J.: Normative models of judgment and decision making. In: Koehler, D.J., Harvey, N. (eds.) Blackwell Handbook of Judgment and Decision Making, pp. 19–36. Blackwell, London (2004)
21. Bayrakdaroğlu, A., Yalçin, N.: A fuzzy multi criteria evaluation of the operational risk factors for the state-owned and privately-owned commercial banks in Turkey. Human Ecol. Risk Assess. Int. J. **19**(2), 443–461 (2013)
22. Bell, D.E., Raiffa, H., Tversky, A.: Decision making: Descriptive, normative and prescriptive interactions. Cambridge University Press, Cambridge (1988)
23. Bellotti, T., Crook, J.: Support vector machines for credit scoring and discovery of significant features. Expert Syst. Appl. **36**(2), 3302–3308 (2009)
24. Belton, V., Stewart, T.J.: Multiple Criteria Decision Analysis: An Integrated Approach. Kluwer, Dordrecht (2002)
25. Abdelaziz, F.B., Aouni, B., El Fayedh, R.: Multi-objective stochastic programming for portfolio selection. Eur. J. Oper. Res. **177**(3), 1811–1823 (2007)
26. Bhaskar, K., McNamee, P.: Multiple objectives in accounting and finance. J. Bus. Finance Account. **10**(4), 595–621 (1983)
27. Bilbao-Terol, A., Arenas-Parra, M., Cañal Fernández, V.: Selection of socially responsible portfolios using goal programming and fuzzy technology. Inf. Sci. **189**, 110–125 (2012)
28. Bilbao-Terol, A., Arenas-Parra, M., Cañal Fernández, V., Bilbao-Terol, C.: Selection of socially responsible portfolios using hedonic prices. J. Bus. Ethics **115**(3), 515–529 (2013)
29. Black, F., Scholes, M.: The pricing of options and corporate liabilities. J. Political Econ. **81**(3), 637–659 (1973)
30. Blöchlinger, A., Leippold, M.: Economic benefit of powerful credit scoring. J. Bank. Finance **30**(3), 851–873 (2006)
31. Bouyssou, D.: Building criteria: a prerequisite for MCDA. In: Bana e Costa, C.A. (ed.) Readings in Multiple Criteria Decision Aid, pp. 27–71. Springer, Berlin (1990)
32. Bouyssou, D.: Using DEA as a tool for MCDM: some remarks. J. Oper. Res. Society **50**, 974–978 (1999)
33. Vincke, J.P., Brans, P.: A preference ranking organization method. Manage. Sci. **31**(6), 647–656 (1985)
34. Brealey, R.A., Myers, S.C., Allen, F.: Principles of Corporate Finance, 10th edn. McGraw-Hill, New York (2011)
35. Briec, W., Kerstens, K.: Portfolio selection in multidimensional general and partial moment space. J. Econ. Dyn. Control **34**(4), 636–656 (2010)
36. Briec, W., Kerstens, K., Jokund, O.: Mean-variance-skewness portfolio performance gauging: a general shortage function and dual approach. Manage. Sci. **53**, 135–149 (2007)
37. Bugera, V., Konno, H., Uryasev, S.: Credit cards scoring with quadratic utility function. J. Multi-Criteria Decis. Anal. **11**(4–5), 197–211 (2002)
38. Calverley, J.: Country Risk Analysis, 2nd edn. Butterworth and Co Publishers Ltd., Scotland (1990)
39. Capotorti, A., Barbanera, E.: Credit scoring analysis using a fuzzy probabilistic rough set model. Comput. Stat. Data Anal. **56**(4), 981–994 (2012)
40. Carhart, M.: On persistence in mutual fund performance. J. Finance **52**(1), 57–82 (1997)

41. Chan, Y., DiSalvo, J.P., Garrambone, M.W.: A goal-seeking approach to capital budgeting. Socio-Economic Plan. Sci. **39**(2), 165–182 (2005)
42. Chang, C.C., Lin, C.J.: LIBSVM: a library for support vector machines. ACM Trans. Intell. Syst. Technol. **2**, 27:1–27:27 (2011)
43. Charnes, A., Cooper, W.W., Rhodes, E.: Measuring the efficiency of decision making units. Eur. J. Oper. Res. **2**, 429–444 (1978)
44. Chiam, S., Tan, K., Mamun, A.A.: Dynamic index tracking via multi-objective evolutionary algorithm. Appl. Soft Comput. **13**(7), 3392–3408 (2013)
45. Erb, C.B., Harvey, C.R., Viskanta, T.E.: Country Risk in Global Financial Management. Research Foundation Publications, Charlottesville (1998)
46. Coelli, T.J., Rao, P.D.S., O'Donnell, C.J., Battese, G.E.: An Introduction to Efficiency and Productivity Analysis, 2nd edn. Springer, New York (2005)
47. Coleman, L.: Why Managers and Companies Take Risks. Springer, London (2006)
48. Comptroller of the Currency Administrator of National Banks: Rating Credit Risk: Comptroller's Handbook. U.S. Department of Treasury (2001)
49. Cornuejols, G., Tutuncu, R.: Optimization Methods in Finance. Cambridge University Press, Cambridge (2007)
50. Cosset, J.C., Siskos, Y., Zopounidis, C.: Evaluating country risk: a decision support approach. Glob. Finance J. **3**(1), 79–95 (1992)
51. e Costa, C.A.B., Barroso, L., Soares, J.O.: Qualitative modelling of credit scoring: a case study in banking. J. Eur. Res. Stud. **5**(1–2), 37–51 (2002)
52. e Costa, C.A.B., Soares, J.O.: A multicriteria model for portfolio management. Eur. J. Finance **10**, 198–211 (2004)
53. Courtis, J.K.: Modelling a financial ratios categoric framework. J. Bus. Finance Account. **5**(4), 371–386 (1978)
54. Crook, J.N., Edelman, D.B., Thomas, L.C.: Recent developments in consumer credit risk assessment. Eur. J. Oper. Res. **183**(3), 1447–1465 (2007)
55. Cundrič, A., Kern, T., Rajkovič, V.: A qualitative model for road investment appraisal. Transp. Policy **15**(4), 225–231 (2008)
56. Dahl, H., Meeraus, A., Zenios, S.A.: Some financial optimization models: I risk management. In: Dahl, H., Meeraus, A., Zenios, S.A. (eds.) Financial Optimization, pp. 3–36. Cambridge University Press, Cambridge (1993)
57. Dash, G.H., Kajiji, N.: Evolving economy bank asset-liability and risk management under uncertainty with hierarchical objectives and nonlinear pricing. J. Multi-Criteria Decis. Anal. **11**(4–5), 247–260 (2002)
58. Davies, R.J., Kat, H.M., Lu, S.: Fund of hedge funds portfolio selection: a multiple-objective approach. J. Deriv. Hedge Funds **15**, 91–115 (2009)
59. Deb, K., Steuer, R.E., Tewari, R., Tewari, R.: Bi-objective portfolio optimization using a customized hybrid NSGA-II procedure. In: Takahashi, R.H.C., Deb, K., Wanner, E.F., Greco, S. (eds.) Evolutionary Multi-Criterion Optimization. Lecture Notes in Computer Science, vol. 6576, pp. 358–373. Springer, Berlin (2011)
60. Delimatsis, P.: Transparent financial innovation in a post-crisis environment. J. Int. Econ. Law **16**(1), 159–210 (2013)
61. Dimitras, A., Slowinski, R., Susmaga, R., Zopounidis, C.: Business failure prediction using rough sets. Eur. J. Oper. Res. **114**(2), 263–280 (1999)
62. Dixon, R.: Venture capitalists and the appraisal of investments. Omega **19**(5), 333–344 (1991)
63. Doumpos, M.: Learning non-monotonic additive value functions for multicriteria decision making. OR Spectrum **34**(1), 89–106 (2012)
64. Doumpos, M., Grigoroudis, E.: Multicriteria Decision Aid and Artificial Intelligence: Links, Theory and Applications. Wiley, New York (2013)
65. Doumpos, M., Kosmidou, K., Baourakis, G., Zopounidis, C.: Credit risk assessment using a multicriteria hierarchical discrimination approach: a comparative analysis. Eur. J. Oper. Res. **138**(2), 392–412 (2002)

66. Doumpos, M., Kosmidou, K., Zopounidis, C.: Analysis and Evaluation of Country Risk: Theoretical and Empirical Approach. Klidarithmos, Athens (2004). (in Greek)
67. Doumpos, M., Marinakis, Y., Marinaki, M., Zopounidis, C.: An evolutionary approach to construction of outranking models for multicriteria classification: the case of the ELECTRE TRI method. Eur. J. Oper. Res. **199**(2), 496–505 (2009)
68. Doumpos, M., Niklis, D., Zopounidis, C., Andriosopoulos, K.: Combining accounting data and a structural model for predicting credit ratings: empirical evidence from european listed firms. J. Bank. Finance (2014). doi:10.1016/j.jbankfin.2014.01.010
69. Doumpos, M., Pasiouras, F.: Developing and testing models for replicating credit ratings: a multicriteria approach. Comput. Econ. **25**(4), 327–341 (2005)
70. Doumpos, M., Zopounidis, C.: Multicriteria Decision Aid Classification Methods. Springer, New York (2002)
71. Doumpos, M., Zopounidis, C.: Regularized estimation for preference disaggregation in multiple criteria decision making. Comput. Optim. Appl. **38**(1), 61–80 (2007)
72. Doumpos, M., Zopounidis, C.: Monotonic support vector machines for credit rating. New Math. Nat. Comput. **5**(3), 557–570 (2009)
73. Doumpos, M., Zopounidis, C.: A multicriteria decision support system for bank rating. Decis. Support Syst. **50**(1), 55–63 (2010)
74. Doumpos, M., Zopounidis, C.: A multicriteria outranking modeling approach for credit rating. Decis. Sci. **42**(3), 721–742 (2011)
75. Doumpos, M., Zopounidis, C.: The robustness concern in preference disaggregation approaches for decision aiding: an overview. In: Rassias, T., Floudas, C., Butenko, S. (eds.) Optimization in Science & Engineering, pp. 161–182. Springer, Berlin (2014)
76. Doumpos, M., Zopounidis, C., Pardalos, P.M.: Financial Decision Making Using Computational Intelligence. Springer, New York (2012)
77. Drake, L., Hall, M.J.B., Simper, R.: The impact of macroeconomic and regulatory factors on bank efficiency: a non-parametric analysis of Hong Kong's banking system. J. Bank. Finance **30**, 1443–1466 (2006)
78. Dyer, J.S.: MAUT—Multiattribute utility theory. In: Figueira, J.R., Greco, S., Ehrgott, M. (eds.) Multiple Criteria Decision Analysis: State of the Art Surveys, pp. 265–285. Springer, Boston (2005)
79. Efron, B., Tibshirani, R.: An Introduction to the Bootstrap. Chapman & Hall, New York (1993)
80. Ehrgott, M., Figueira, J.R., Greco, S.: Trends in Multiple Criteria Decision Analysis. Springer, New York (2010)
81. Ehrgott, M., Klamroth, K., Schwehm, C.: An MCDM approach to portfolio optimization. Eur. J. Oper. Res. **155**(3), 752–770 (2004)
82. Englemann, B., Hayden, E., Tasche, D.: Testing rating accuracy. Risk **16**, 82–86 (2003)
83. Fabozzi, F.J., Focardi, S., Jonas, C.: Trends in quantitative equity management: survey results. Quant. Financ. **7**(2), 115–122 (2007)
84. Fabozzi, F.J., Peterson, P.P.: Financial Management and Analysis. Wiley, Hoboken (2003)
85. Fama, E.F., French, K.R.: Common risk factors in the returns on stocks and bonds. J. Financ Econ **33**(1), 3–56 (1993)
86. Fama, E.F., French, K.R.: Multifactor explanations of asset pricing anomalies. J. Finance **51**(1), 55–84 (1996)
87. Fawcett, T.: An introduction to ROC analysis. Pattern Recogn. Lett. **27**(8), 861–874 (2006)
88. Fethi, M.D., Pasiouras, F.: Assessing bank efficiency and performance with operational research and artificial intelligence techniques: a survey. Eur. J. Oper. Res. **204**(2), 189–198 (2010)
89. Figueira, J.R., Greco, S., Roy, B., Slowinski, R.: ELECTRE methods: main features and recent developments. In: Zopounidis, C., Pardalos, P.M. (eds.) Handbook of Multicriteria Analysis, pp. 51–89. Springer, Berlin (2010)
90. Finnerty, J.D.: Financial engineering in corporate finance: an overview. Financ. Manage. **17**(4), 14–33 (1988)

91. Fisk, D.: The 2008 financial collapse: lessons for engineering failure. Eng. Fail. Anal. **18**(2), 550–556 (2011)
92. Friedman, C., Sandow, S.: Model performance measures for expected utility maximizing investors. Int. J. Theor. Appl. Finance **6**(4), 355–401 (2003)
93. Frost, C.A.: Credit rating agencies in capital markets: a review of research evidence on selected criticisms of the agencies. J. Account. Auditing Finance **22**(3), 469–492 (2007)
94. Fund, I.M.: Global financial stability report—Sovereigns, funding, and systemic liquidity, Washington, DC. (2010)
95. Gaganis, C., Pasiouras, F., Doumpos, M., Zopounidis, C.: Modelling banking sector stability with multicriteria approaches. Optim. Lett. **4**, 543–558 (2010)
96. Gaganis, C., Pasiouras, F., Zopounidis, C.: A multicriteria decision framework for measuring banks' soundness around the world. J. Multi-Criteria Decis. Anal. **14**(1–3), 103–111 (2006)
97. Gaivoronski, A.A., Krylov, S., van der Wijst, N.: Optimal portfolio selection and dynamic benchmark tracking. Eur. J. Oper. Res. **163**(1), 115–131 (2005)
98. Gaivoronski, A.A., Pflug, G.: Value-at-risk in portfolio optimization: properties and computational approach. J. Risk **7**(2), 1–31 (2005)
99. García, F., Giménez, V., Guijarro, F.: Credit risk management: a multicriteria approach to assess creditworthiness. Math. Comput. Model. **57**(7–8), 2009–2015 (2013)
100. García, F., Guijarro, F., Moya, I.: A goal programming approach to estimating performance weights for ranking firms. Comput. Oper. Res. **37**(9), 1597–1609 (2010)
101. Gilli, M., Maringer, D., Winker, P.: Applications of heuristics in finance. In: Seese, D., Weinhardt, C., Schlottmann, F. (eds.) Handbook on Information Technology in Finance, pp. 635–653. Springer, Berlin (2008)
102. Glasserman, P.: Monte Carlo Methods in Financial Engineering. Springer, New York (2003)
103. Götze, U., Northcott, D., Schuster, P.: Investment Appraisal—Methods and Models. Springer, Berlin (2008)
104. Grabisch, M., Kojadinovic, I., Meyer, P.: A review of methods for capacity identification in Choquet integral based multi-attribute utility theory: applications of the Kappalab R package. Eur. J. Oper. Res. **186**(2), 766–785 (2008)
105. Greco, S., Matarazzo, B., Slowinski, R.: Rough sets theory for multicriteria decision analysis. Eur. J. Oper. Res. **129**(1), 1–47 (2001)
106. Grigoroudis, E., Tsitsiridi, E., Zopounidis, C.: Linking customer satisfaction, employee appraisal, and business performance: an evaluation methodology in the banking sector. Ann. Oper. Res. **205**(1), 5–27 (2013)
107. Grunert, J., Norden, L., Weber, M.: The role of non-financial factors in internal credit ratings. J. Bank. Finance **29**(2), 509–531 (2005)
108. van Haaren, R., Fthenakis, V.: GIS-based wind farm site selection using spatial multi-criteria analysis (smca): evaluating the case for New York State. Renew. Sustain. Energy Rev. **15**(7), 3332–3340 (2011)
109. Hall, J., Hofer, C.W.: Venture capitalists' decision criteria in new venture evaluation. J. Bus. Ventur. **8**(1), 25–42 (1993)
110. Hallerbach, W., Ning, H., Soppe, A., Spronk, J.: A framework for managing a portfolio of socially responsible investments. Eur. J. Oper. Res. **153**(2), 517–529 (2004)
111. Hallerbach, W.G., Spronk, J.: The relevance of MCDM for financial decisions. J. Multi-Criteria Decis. Anal. **11**(4–5), 187–195 (2002)
112. Han, S.H., Diekmann, J.E., Lee, Y., Ock, J.H.: Multicriteria financial portfolio risk management for international projects. J. Constr. Eng. Manage. **130**(3), 346–356 (2004)
113. He, J., Zhang, Y., Shi, Y., Huang, G.: Domain-driven classification based on multiple criteria and multiple constraint-level programming for intelligent credit scoring. IEEE Trans. Knowl. Data Eng. **22**(6), 826–838 (2010)
114. Henriksson, R.D., Merton, R.C.: On market timing and investment performance ii: statistical procedures for evaluating forecasting skills. J. Bus. **54**(4), 513 (1981)
115. Herring, R.: Managing International Risk. Cambridge University Press, Cambridge (1983)

116. Hirschberger, M., Qi, Y., Steuer, R.E.: Large-scale MV efficient frontier computation via a procedure of parametric quadratic programming. Eur. J. Oper. Res. **204**(3), 581–588 (2010)
117. Ho, C.T.: Measuring bank operations performance: an approach based on grey relation analysis. J. Oper. Res. Society **57**(4), 337–349 (2006)
118. Horcher, K.A.: Essentials of of Financial Risk Management. Wiley, Hoboken (2005)
119. Hu, J.L., Chang, T.P.: Decomposition of mutual fund underperformance. Appl. Financ. Econ. Lett. **4**(5), 363–367 (2008)
120. Hu, Y.C.: Bankruptcy prediction using ELECTRE-based single-layer perceptron. Neurocomputing **72**, 3150–3157 (2009)
121. Hu, Y.C., Chen, C.J.: A PROMETHEE-based classification method using concordance and discordance relations and its application to bankruptcy prediction. Inf. Sci. **181**(22), 4959–4968 (2011)
122. Hu, Y.C., Liao, P.C.: Finding critical criteria of evaluating electronic service quality of internet banking using fuzzy multiple-criteria decision making. Appl. Soft Comput. **11**(4), 3764–3770 (2011)
123. Huang, S.C.: Using Gaussian process based kernel classifiers for credit rating forecasting. Expert Syst. Appl. **38**(7), 8607–8611 (2011)
124. Hurson, C., Zopounidis, C.: On the use of multicriteria decision aid methods to portfolio selection. In: Clímaco, J. (ed.) Multicriteria Analysis, pp. 496–507. Springer, Berlin (1997)
125. Ioannidis, C., Pasiouras, F., Zopounidis, C.: Assessing bank soundness with classification techniques. Omega **38**(5), 345–357 (2010)
126. Jablonsky, J.: Multicriteria approaches for ranking of efficient units in DEA models. Central Eur. J. Oper. Res. **20**, 435–449 (2012)
127. Jacquet-Lagrèze, E., Siskos, Y.: Assessing a set of additive utility functions for multicriteria decision making: the UTA method. Eur. J. Oper. Res. **10**(2), 151–164 (1982)
128. Jacquet-Lagrèze, E., Siskos, Y.: Preference disaggregation: twenty years of MCDA experience. Eur. J. Oper. Res. **130**(2), 233–245 (2001)
129. Jensen, M.C.: The performance of mutual funds in the period 1945–1964. J. Finance **23**(2), 389–416 (1968)
130. Jeon, D.S., Lovo, S.: Credit rating industry: a helicopter tour of stylized facts and recent theories. Int. J. Ind. Organ. **31**(5), 643–651 (2013)
131. Jia, J., Fischer, G., Dyer, J.: Attribute weighting methods and decision quality in the presence of response error: a simulation study. J. Behav. Decis. Mak. **11**, 85–105 (1998)
132. Jiménez, L., Pascual, L.: Multicriteria cash-flow modeling and project value-multiples for two-stage project valuation. Int. J. Project Manage. **26**(2), 185–194 (2008)
133. Jondeau, W., Rockinger, M.: Optimal portfolio allocation under higher moments. Eur. Financ. Manage. **12**(1), 29–55 (2006)
134. Jones, D., Tamiz, M.: Practical Goal Programming. Springer, London (2012)
135. Jorion, P.: Value At Risk, 3rd edn. McGraw-Hill, New York (2009)
136. Joro, T., Korhonen, P., Wallenius, J.: Structural comparison of data envelopment analysis and multiple objective linear programming. Manage. Sci. **44**(7), 962–970 (1998)
137. Kahneman, D., Tversky, A.: Prospect theory: an analysis of decision under risk. Econometrica **47**(2), 263 (1979)
138. Keeney, R.L.: Common mistakes in making value trade-offs. Oper. Res. **50**(6), 935–945 (2002)
139. Keeney, R.L., Raiffa, H.: Decisions with Multiple Objectives: Preferences and Value Trade-offs. Cambridge University Press, Cambridge (1993)
140. Kerstens, K., Mounir, A., van de Woestyne, I.: Geometric representation of the mean-variance-skewness portfolio frontier based upon the shortage function. Eur. J. Oper. Res. **210**(1), 81–94 (2011)
141. Kerstens, K., Mounir, A., de Woestyne, I.V.: Non-parametric frontier estimates of mutual fund performance using C- and L-moments: some specification tests. J. Bank. Finance **35**(5), 1190–1201 (2011)
142. Kolm, P., Tütüncü, R., Fabozzi, F.: 60 years of portfolio optimization: practical challenges and current trends. Eur. J. Oper. Res. **234**(2), 356–371 (2014)

143. Konno, H., Yamazaki, H.: Mean-absolute deviation portfolio optimization model and its applications to tokyo stock market. Manage. Sci. **37**(5), 519–531 (1991)
144. Kosmidou, K., Doumpos, M., Zopounidis, C.: Country Risk Evaluation: Methods and Applications. Springer, New York (2008)
145. Kosmidou, K., Zopounidis, C.: Goal Programming Techniques for Bank Asset Liability Management. Springer, Boston (2004)
146. Kosmidou, K., Zopounidis, C.: Measurement of bank performance in Greece. South-Eastern Eur. J. Econ. **6**(1), 79–95 (2008)
147. Krahnen, J.P., Weber, M.: Generally accepted rating principles: a primer. J. Bank. Finance **25**(1), 3–23 (2001)
148. Krink, T., Paterlini, S.: Multiobjective optimization using differential evolution for real-world portfolio optimization. Comput. Manage. Sci. **8**, 157–179 (2011)
149. Kunsch, P.: Advanced operations research techniques in capital budgeting. In: Zopounidis, C., Doumpos, M., Pardalos, P. (eds.) Handbook of Financial Engineering, pp. 301–342. Springer, New York (2008)
150. Kuosmanen, T.: Performance measurement and best-practice benchmarking of mutual funds: combining stochastic dominance criteria with data envelopment analysis. J. Prod. Anal. **28**(1–2), 71–86 (2007)
151. Lahdelma, R., Salminen, P.: SMAA-2: stochastic multicriteria acceptability analysis for group decision making. Oper. Res. **49**(3), 444–454 (2001)
152. Lai, T.Y.: Portfolio selection with skewness: a multiple-objective approach. Rev. Quant. Financ. Account. **1**, 293–305 (1991)
153. Lerner, J.: The new new financial thing: the origins of financial innovations. J. Financ. Econ **79**(2), 223–255 (2006)
154. Li, J., Wei, L., Li, G., Xu, W.: An evolution strategy-based multiple kernels multi-criteria programming approach: the case of credit decision making. Decis. Supp. Syst. **51**(2), 292–298 (2011)
155. Lo, A., Petrov, C., Wierzbicki, M.: Its 11pm—Do you know where your liquidity is? The mean-variance-liquidity frontier. J. Invest. Manage. **1**(1), 55–93 (2003)
156. Lozano, S., Gutiérrez, E.: Data envelopment analysis of mutual funds based on second-order stochastic dominance. Eur. J. Oper. Res. **189**(1), 230–244 (2008)
157. Lyra, M., Paha, J., Paterlini, S., Winker, P.: Optimization heuristics for determining internal rating grading scales. Comput. Stat. Data Anal. **54**(11), 2693–2706 (2010)
158. Macharis, C., de Witte, A., Ampe, J.: The multi-actor, multi-criteria analysis methodology (MAMCA) for the evaluation of transport projects: theory and practice. J. Adv. Transp. **43**(2), 183–202 (2009)
159. Macmillan, I.C., Siegel, R., Narasimha, P.: Criteria used by venture capitalists to evaluate new venture proposals. J. Bus. Ventur. **1**(1), 119–128 (1985)
160. Maginn, J., Tuttle, D., McLeavey, D., Pinto, J.: Managing Investment Portfolios—A Dynamic Process. Wiley, New York (2007)
161. Mansini, R., Ogryczak, W., Speranza, M.G.: Conditional value at risk and related linear programming models for portfolio optimization. Ann. Oper. Res. **152**, 227–256 (2007)
162. Mansini, R., Ogryczak, W., Speranza, M.G.: Twenty years of linear programming based portfolio optimization. Eur. J. Oper. Res. **234**(2), 518–535 (2014)
163. Mareschal, B., Brans, J.: BANKADVISER: an industrial evaluation system. Eur. J. Oper. Res. **54**, 318–324 (1991)
164. Mareschal, B., Mertens, D.: BANKS: a multicriteria decision support system for financial evaluation in the international banking sector. J. Decis. Syst. **1**(2–3), 175–189 (1992)
165. Markowitz, H.: Mean-variance approximations to expected utility. Eur. J. Oper. Res. **234**(2), 346–355 (2014)
166. Markowitz, H., Todd, P., Xu, G., Yamane, Y.: Computation of mean-semivariance efficient sets by the critical line algorithm. Ann. Oper. Res. **45**(1), 307–317 (1993)
167. Markowitz, H.M.: Portfolio Selection: Efficient Diversification of Investment. Wiley, New York (1959)

168. Marshall, J.F., Dorigan, M.P.: Financial engineering: information technology and its place in the new finance. Technol. Soc. **18**(2), 185–201 (1996)
169. Martel, J.M., Matarazzo, B.: Other outranking approaches. In: Multiple Criteria Decision Analysis: State of the Art Surveys, pp. 197–259. Springer, New York (2005)
170. Martens, D., Baesens, B.: Building acceptable classification models. In: Stahlbock, R., Crone, S.F., Lessmann, S. (eds.) Data Mining, pp. 53–74. Springer, New York (2010)
171. Martens, D., Baesens, B., Van Gestel, T., Vanthienen, J.: Comprehensible credit scoring models using rule extraction from support vector machines. Eur. J. Oper. Res. **183**(3), 1466–1476 (2007)
172. Martens, D., Vanthienen, J., Verbeke, W., Baesens, B.: Performance of classification models from a user perspective. Decis. Supp. Syst. **51**(4), 782–793 (2011)
173. Matallín-Sáez, J., Soler-Domínguez, A., Tortosa-Ausina, E.: On the informativeness of persistence for evaluating mutual fund performance using partial frontiers. Omega **42**(1), 47–64 (2014)
174. Mavrotas, G.: Effective implementation of the ε-constraint method in multi-objective mathematical programming problems. Appl. Math. Comput. **213**(2), 455–465 (2009)
175. Metaxiotis, K., Liagkouras, K.: Multiobjective evolutionary algorithms for portfolio management: a comprehensive literature review. Expert Syst. Appl. **39**(14), 11685–11698 (2012)
176. Michalowski, W., Ogryczak, W.: Extending the MAD portfolio optimization model to incorporate downside risk aversion. Naval Res. Logistics **48**(3), 185–200 (2001)
177. Michaud, R., Michaud, R.: Efficient Asset Management—A Practical Guide to Stock Portfolio Optimization and Asset Allocation, 2nd edn. Oxford University Press, New York (2008)
178. Michaud, R.O.: The markowitz optimization enigma: Is 'optimized' optimal? Financ. Anal. J. **45**(1), 31–42 (1989)
179. Miettinen, K.: Nonlinear Multiobjective Optimization. Kluwer Academic Publishers, Dordrecht (1999)
180. Montibeller, G., Franco, L.A., Lord, E., Iglesias, A.: Structuring resource allocation decisions: a framework for building multi-criteria portfolio models with area-grouped options. Eur. J. Oper. Res. **199**(3), 846–856 (2009)
181. Morton, G., Johnathan, M.: Credit Engineering for Bankers: A Practical Guide for Bank Lending, 2nd edn. Academic Press, Boston (2011)
182. Mousseau, V., Slowinski, R.: Inferring an ELECTRE TRI model from assignment examples. J. Global Optim. **12**(2), 157–174 (1998)
183. Mulvey, J.M., Rosenbaum, D.P., Shetty, B.: Strategic financial risk management and operations research. Eur. J. Oper. Res. **97**(1), 1–16 (1997)
184. Muzyka, D., Birley, S., Leleux, B.: Trade-offs in the investment decisons of european venture capitalists. J. Bus. Ventur. **11**(4), 273–287 (1996)
185. von Neumann, J., Morgenstern, O.: Theory of Games and Economic Behavior. Princeton University Press, Princeton (1944)
186. Odeh, O., Koduru, P., Featherstone, A., Das, S., Welch, S.M.: A multi-objective approach for the prediction of loan defaults. Expert Syst. Appl. **38**(7), 8850–8857 (2011)
187. Ogryczak, W.: Multiple criteria linear programming model for portfolio selection. Ann. Oper. Res. **97**, 143–162 (2000)
188. Oliver, R.M.: Financial performance measures in credit scoring. Eur. J. Decis. Processes **1**(3–4), 169–185 (2013)
189. Oral, M., Kettani, O., Cosset, J.C., Daouas, M.: An estimation model for country risk rating. Int. J. Forecast. **8**(4), 583–593 (1992)
190. Pagano, M., Volpin, P.: Credit ratings failures and policy options. Econ. Policy **25**(62), 401–431 (2010)
191. Pahikkala, T., Waegeman, W., Tsivtsivadze, E., Salakoski, T., De Baets, B.: Learning intransitive reciprocal relations with kernel methods. Eur. J. Oper. Res. **206**(3), 676–685 (2010)
192. Papadopoulos, A., Karagiannidis, A.: Application of the multi-criteria analysis method ELECTRE III for the optimisation of decentralised energy systems. Omega **36**(5), 766–776 (2008)

193. Papageorgiou, D., Doumpos, M., Zopounidis, C., Pardalos, P.M.: Credit rating systems: Regulatory framework and comparative evaluation of existing methods. In: Zopounidis, C., Doumpos, M., Pardalos, P.M. (eds.) Handbook of Financial Engineering, pp. 457–488. Springer, New York (2008)
194. Pardalos, P.M., Siskos, Y., Zopounidis, C.: Advances in Multicriteria Analysis. Kluwer Academic Publishers, Dordrecht (1995)
195. Pasiouras, F.: Estimating the technical and scale efficiency of Greek commercial banks: The impact of credit risk, off-balance sheet activities, and international operations. Res. Int. Bus. Finance **22**, 301–318 (2008)
196. Pendaraki, K., Zopounidis, C., Doumpos, M.: On the construction of mutual fund portfolios: a multicriteria methodology and an application to the Greek market of equity mutual funds. Eur. J. Oper. Res. **163**(2), 462–481 (2005)
197. Peng, Y., Kou, G., Shi, Y., Chen, Z.: A multi-criteria convex quadratic programming model for credit data analysis. Decis. Supp. Syst. **44**(4), 1016–1030 (2008)
198. Peón, D., Antelo, M.: Are normative models in finance realistic? Cuad. Economía **35**(98), 89–99 (2012)
199. Premachandra, I., Zhu, J., Watson, J., Galagedera, D.U.: Best-performing US mutual fund families from 1993 to 2008: evidence from a novel two-stage DEA model for efficiency decomposition. J. Bank. Finance **36**(12), 3302–3317 (2012)
200. Raveh, A.: The Greek banking system: reanalysis of performance. Eur. J. Oper. Res. **120**(3), 525–534 (2000)
201. Rigopoulos, G., Anagnostopoulos, K.: Fuzzy multicriteria assignment for nominal classification: methodology and application in evaluation of greek banks electronic payment retailers. Int. J. Inf. Technol. Decis. Making **9**(3), 437–454 (2010)
202. Rockafellar, R.T., Uryasev, S.: Optimization of conditional value-at-risk. J. Risk **2**, 21–41 (2000)
203. Rodríguez, R., Luque, M., González, M.: Portfolio selection in the Spanish stock market by interactive multiobjective programming. TOP **19**, 213–231 (2011)
204. Roman, D., Darby-Dowman, K., Mitra, G.: Mean-risk models using two risk measures: a multiobjective approach. Quant. Finance **7**(4), 443–458 (2007)
205. Ross, S.A.: The arbitrage theory of capital asset pricing. J. Econ. Theory **13**(3), 341–360 (1976)
206. Rousos, E.P., Lee, B.S.: Multicriteria analysis in shipping investment evaluation. Marit. Policy Manage. **39**(4), 423–442 (2012)
207. Roy: Multicriteria Methodology for Decision Aiding. Kluwer Academic Publisher, Dordrecht (1996)
208. Roy, B.: Classement et choix en présence de points de vue multiples: La méthode ELECTRE. Revue Francaise d'Informatique et de Recherche Opérationnelle **8**, 57–75 (1968)
209. Roy, B., Bouyssou, D.: Aide Multicritère à la Décision: Méthodes et Cas. Economica, Paris (1993)
210. Ruppert, D.: Statistics and Finance. Springer, New York (2004)
211. Saaty, T.L., Rogers, P.C., Pell, R.: Portfolio selection through hierarchies. J. Portfolio Manage. **6**(3), 16–21 (1980)
212. Sahajwala, R., Van den Bergh, P.: Supervisory risk assessment and early warning systems. Technical Report 4. Bank of International Settlements, Basel, Switzerland (2000)
213. Samaras, G.D., Matsatsinis, N.F., Zopounidis, C.: A multicriteria DSS for stock evaluation using fundamental analysis. Eur. J. Oper. Res. **187**(3), 1380–1401 (2008)
214. Sarkis, J.: A comparative analysis of DEA as a discrete alternative multiple criteria decision tool. Eur. J. Oper. Res. **123**(3), 543–557 (2000)
215. Sharpe: Expected utility asset allocation. Financ. Anal. J. **63**(5), 18–30 (2007)
216. Sharpe, W.F.: Capital asset prices: a theory of market equilibrium under conditions of risk. J. Finance **19**(3), 425–442 (1964)
217. Sharpe, W.F.: Mutual fund performance. J. Bus. **39**(S1), 119–138 (1966)
218. Shreve, S.: Stochastic Calculus for Finance, vol. I, II. Springer, New York (2004)

219. Siddiqi, N.: Credit risk scorecards. Wiley, Hoboken (2006)
220. Siskos, Y., Grigoroudis, E.: New trends in aggregation-disaggregation approaches. In: Zopounidis, C., Pardalos, P. (eds.) Handbook of Multicriteria Analysis, pp. 189–214. Springer, Berlin (2010)
221. Siskos, Y., Yannacopoulos, D.: UTASTAR: An ordinal regression method for building additive value functions. Investigação Operacional **5**(1), 39–53 (1985)
222. Sobehart, J., Keenan, S.: Measuring default accurately. Risk **14**(3), 31–33 (2001)
223. Spathis, C., Kosmidou, K., Doumpos, M.: Assessing profitability factors in the Greek banking system: a multicriteria approach. Int. Trans. Oper. Res. **9**, 517–530 (2002)
224. Spronk, J., Hallerbach, W.: Financial modelling: Where to go? With an illustration for portfolio management. Eur. J. Oper. Res. **99**(1), 113–125 (1997)
225. Spronk, J., Steuer, R.E., Zopounidis, C.: Multicriteria decision aid/analysis in finance. In: Figueira, J., Greco, S., Ehrgott, M. (eds.) Multiple Criteria Decision Analysis: State of the Art Surveys, pp. 799–858. Springer, Boston (2005)
226. Srinivasan, V., Shocker, A.D.: Linear programming techniques for multidimensional analysis of preferences. Psychometrika **38**, 337–369 (1973)
227. Stankevičiene, J., Mencaite, E.: The evaluation of bank performance using a multicriteria decision making model: a case study on Lithuanian commercial banks. Technol. Econ. Dev. Econ. **18**(1), 189–205 (2012)
228. Stein, R.M.: Benchmarking default prediction models: Pitfalls and remedies in model validation. J. Risk Model Validation **1**(1), 77–113 (2007)
229. Steuer, R.E.: Multiple Criteria Optimization: Theory, Computation and Application. Wiley, New York (1986)
230. Steuer, R.E., Na, P.: Multiple criteria decision making combined with finance: a categorized bibliographic study. Eur. J. Oper. Res. **150**(3), 496–515 (2003)
231. Steuer, R.E., Qi, Y., Hirschberger, M.: Suitable-portfolio investors, nondominated frontier sensitivity, and the effect of multiple objectives on standard portfolio selection. Ann. Oper. Res. **152**, 297–317 (2007)
232. Steuer, R.E., Qi, Y., Hirschberger, M.: Portfolio selection in the presence of multiple criteria. In: Zopounidis, C., Doumpos, M., Pardalos, P.M. (eds.) Handbook of Financial Engineering, pp. 3–24. Springer, Berlin (2008)
233. Stone, M.: Cross-validatory choice and assessment of statistical predictions. J. Royal Stat. Soc. Ser. B **36**(2), 111–147 (1974)
234. Su, C.T., Chen, Y.C.: Rule extraction algorithm from support vector machines and its application to credit screening. Soft Comput. **16**(4), 645–658 (2012)
235. Szegö, G.: Risk Measures for the 21st Century. Wiley, New York (2004)
236. Szegö, G.: Measures of risk. J. Bank. Finance **26**, 1253–1272 (2005)
237. Tang, J., Espinal, C.: A model to assess country risk. Omega **17**(4), 363–367 (1989)
238. Tektas, A., Ozkan-Gunay, E.N., Gunay, G.: Asset and liability management in financial crisis. J. Risk Finance **6**(2), 135–149 (2005)
239. Tervonen, T., Figueira, J.R.: A survey on stochastic multicriteria acceptability analysis methods. J. Multi-Criteria Decis. Anal. **15**(1–2), 1–14 (2008)
240. Teynor, J.: How to rate management of investment funds. Harvard Bus. Rev. **43**, 63–75 (1965)
241. Tichy, G., Lannoo, K., Gwilym, O., Alsakka, R., Masciandaro, D., Paudyn, B.: Credit rating agencies: Part of the solution or part of the problem? Intereconomics **46**(5), 232–262 (2011)
242. Tikhonov, A.N., Goncharsky, A.V., Stepanov, V.V., Yogala, A.G.: Numerical Methods for the Solution of Ill-Posed Problems. Springer, Dordrecht (1995)
243. Ting, W.: Multinational Risk Assessment and Management. Greenwood Press, Westport (1988)
244. Treacy, W.F., Carey, M.: Credit risk rating systems at large US banks. J. Bank. Finance **24**(1–2), 167–201 (2000)
245. Treynor, J.L., Mazuy, K.: Can mutual funds outguess the market? Harvard Bus. Rev. **44**, 131–136 (1966)

246. Tufano, P.: Financial innovation. In: Constantinides, G.M., Harris, M., Stulz, R.M. (eds.) Handbook of the Economics of Finance, vol. 1, pp. 307–336. Elsevier, Amsterdam (2003)
247. Tuncer Şakar, C., Köksalan, M.: A stochastic programming approach to multicriteria portfolio optimization. J. Global Optim. 1–16 (2012)
248. Utz, S., Wimmer, M., Hirschberger, M., Steuer, R.E.: Tri-criterion inverse portfolio optimization with application to socially responsible mutual funds. Eur. J. Oper. Res. **234**(2), 491–498 (2014)
249. Vapnik, V.: Statistical Learning Theory. Wiley, New York (1998)
250. Vassalou, M., Xing, Y.: Default risk in equity returns. J. Finance **59**(2), 831–868 (2004)
251. Vercher, R., Bermúdez, J.D.: Fuzzy portfolio selection models: a numerical study. In: Doumpos, M., Zopounidis, C., Pardalos, P.M. (eds.) Financial Decision Making Using Computational Intelligence, pp. 253–280. Springer, New York (2012)
252. Vukovic, S., Delibasic, B., Uzelac, A., Suknovic, M.: A case-based reasoning model that uses preference theory functions for credit scoring. Expert Syst. Appl. **39**(9), 8389–8395 (2012)
253. Vučina, D., Lozina, Z., Vlak, F.: NPV-based decision support in multi-objective design using evolutionary algorithms. Eng. Appl. Artif. Intell. **23**(1), 48–60 (2010)
254. Wagner, H.M.: Linear programming techniques for regression analysis. J. Am. Stat. Assoc. **54**, 206–212 (1959)
255. van der Wijst, N.: Finance: A Quantitative Introduction. Cambridge University Press, Cambridge (2013)
256. von Winterfeldt, D., Fasolo, B.: Structuring decision problems: a case study and reflections for practitioners. Eur. J. Oper. Res. **199**(3), 857–866 (2009)
257. Xidonas, P., Mavrotas, G., Krintas, T., Psarras, J., Zopounidis, C.: Multicriteria Portfolio Management. Springer, New York (2012)
258. Xidonas, P., Mavrotas, G., Psarras, J.: A multicriteria methodology for equity selection using financial analysis. Comput. Oper. Res. **36**(12), 3187–3203 (2009)
259. Xidonas, P., Mavrotas, G., Psarras, J.: Equity portfolio construction and selection using multiobjective mathematical programming. J. Global Optim. **47**(2), 185–209 (2010)
260. Xidonas, P., Mavrotas, G., Zopounidis, C., Psarras, J.: IPSSIS: an integrated multicriteria decision support system for equity portfolio construction and selection. Eur. J. Oper. Res. **210**(2), 398–409 (2011)
261. Yu, J.R., Lee, W.Y.: Portfolio rebalancing model using multiple criteria. Eur. J. Oper. Res. **209**(2), 166–175 (2011)
262. Yu, L., Wang, S., Lai, K.: An intelligent-agent-based fuzzy group decision making model for financial multicriteria decision support: the case of credit scoring. Eur. J. Oper. Res. **195**(3), 942–959 (2009)
263. Zhang, J., Shi, Y., Zhang, P.: Several multi-criteria programming methods for classification. Comput. Oper. Res. **36**(3), 823–836 (2009)
264. Zopounidis, C., Despotis, D., Stavropoulou, E.: Multiattribute evaluation of Greek banking performance. Appl. Stoch. Models Data Anal. **11**(1), 97–107 (1995)
265. Zopounidis, C., Doumpos, M.: A multicriteria decision aid methodology for sorting decision problems: the case of financial distress. Comput. Econ. **14**(3), 197–218 (1999)
266. Zopounidis, C., Doumpos, M.: Building additive utilities for multi-group hierarchical discrimination: the M.H.Dis method. Optim. Methods Softw. **14**(3), 219–240 (2000)
267. Zopounidis, C., Doumpos, M.: Intelligent Decision Aiding Systems Based on Multiple Criteria for Financial Engineering. Springer, Dordrecht (2000)
268. Zopounidis, C., Doumpos, M.: INVESTOR: a decision support system based on multiple criteria for portfolio selection and composition. In: Colorni, A., Paruccini, M., Roy, B. (eds.) A-MCD-A (Aide Multi Critere a la Decision—Multiple Criteria Decision Aiding), pp. 371–381. European Commission Joint Research Centre, Brussels (2000)
269. Zopounidis, C., Doumpos, M., Zanakis, S.: Stock evaluation using a preference disaggregation methodology. Decis. Sci. **30**(2), 313–336 (1999)
270. Zopounidis, C., Pardalos, P.M.: Handbook of Multicriteria Analysis. Springer, Berlin (2010)

Index

M. Doumpos and C. Zopounidis, *Multicriteria Analysis in Finance*,
SpringerBriefs in Operations Research, DOI: 10.1007/978-3-319-05864-1,

MIX
Papier aus verantwortungsvollen Quellen
Paper from responsible sources
FSC® C105338

If you have any concerns about our products,
you can contact us on
ProductSafety@springernature.com

In case Publisher is established outside the EU,
the EU authorized representative is:
Springer Nature Customer Service Center GmbH
Europaplatz 3, 69115 Heidelberg, Germany

Printed by Libri Plureos GmbH
in Hamburg, Germany